SOCIAL ECOLOGY

APPLYING ECOLOGICAL UNDERSTANDINGS TO OUR LIVES AND OUR PLANET

Edited by
David Wright,
Catherine E. Camden-Pratt
and Stuart B. Hill

To the universe, it's the only one we have.

Published by Hawthorn Press, Hawthorn House,
1 Lansdown Lane, Stroud, Gloucestershire, GL5 1BJ, UK
Tel: (01453) 757040 Fax: (01453) 751138
E-mail: info@hawthornpress.com
Website: www.hawthornpress.com

Typesetting and design by Bookcraft, Stroud, Gloucestershire, UK
Printed in the UK by Henry Ling Ltd, The Dorset Press, Dorchester.
Printed in Australia by Griffin Press, Clayton, Victoria.

Printed on environmentally friendly paper from renewable forest stock.

British Library Cataloguing in Publication Data applied for

ISBN 978-1-907359-11-8

CONTENTS

LIST OF CONTRIBUTORS

Roslyn Arnold

Roslyn Arnold is an Honorary Professor in the Faculty of Education at the University of Sydney. She was Dean of Education at the University of Tasmania and Pro-Dean and Head of School at the University of Sydney. Her research interests are writing development, arts education, teacher education, leadership development and empathic intelligence – a theory of teaching and learning that she has developed to explain the qualities of exceptional educators. She has delivered keynote addresses on her research interests in Australia, the United States, England, Canada and Singapore. Her books include *Writing Development: Magic in the Brain* (Open University 1992) and *Empathic Intelligence: Teaching, Learning, Relating* (UNSW Press 2005).

Jasmin Ball

Jasmin Ball was raised in Melbourne, Australia. She fell in love with nature at a young age and has always experienced a strong sense of connection to special childhood places. At the age of six she adopted a rock and kept it as a pet, much to the chagrin of her parents! It was this desire for relationship, place-connection, magic and fun that has motivated her to harness diverse and often unexpected opportunities to experience life at its fullest. Jasmin has over ten years experience teaching sustainable living and change in schools, universities, outdoor settings and corporations. With a background in teaching and a master's in social ecology, Jasmin has also lectured and tutored in Education for Sustainability for the University of Western Sydney.

Richard Bawden

Richard Bawden is adjunct professor at Michigan State University, a visiting professor at the Open University in the UK, a Fellow and Director of the Systemic Development Institute (SDI), and a Professor Emeritus at the University of Western Sydney. He has recently retired (August 2007) as a Visiting Distinguished University Professor at Michigan State University. Prior to that he spent 20 years at Hawkesbury Agricultural College, later the University of Western Sydney.

For most of that time he was Dean of Agriculture and Rural Development and Professor of Systemic Development. Upon his retirement from that university he was awarded Professor Emeritus status. He has been a Visiting Scholar/Professor and a consultant to systemic development projects in more than two dozen countries across five continents. He has published more than 200 journal articles, book chapters, and keynote conference papers. He is a member of the editorial boards of three international journals.

Barry Bignell

Barry Bignell studied music at the Royal Military School of Music (UK). He studied conducting at the Royal Academy of Music, and was awarded the Director's Prize for conducting. Barry was subsequently Director of the Australian Army Band, Melbourne. Barry also studied with the Berlin Philharmonic Orchestra and the Deutsche Oper, and subsequently pursued a career as a conductor for orchestras, choirs and windbands. He has had a parallel career in education and was Head of Postgraduate Studies in music at the Victorian College of the Arts and Music, University of Melbourne. He has lectured widely on musicality and its social and psycho-spiritual implications. He is currently preparing a book on that topic. Barry has a master's degree in education from the University of Melbourne and a PhD in social ecology from the University of Western Sydney.

Carol Birrell

Carol is an artist, writer and academic exploring the interaction between an indigenous and Western sense of place. Her 2007 PhD thesis is titled *Meeting Country: Deep Engagement with Place and Indigenous Culture*. She is currently teaching social ecology at the University of Western Sydney and Aboriginal education at the University of Wollongong. She has been working with a land-based arts practice for the last 12 years called 'ecopoiesis', which draws together movement, painting, photography, environmental sculpture and poetry as a base for ecological narratives and exploring our ecological identity. Carol has strong long-term connections with Yuin (south coast NSW) and Worrorra (west Kimberley) indigenous communities.

John Broomfield

Born in New Zealand, John did a doctorate at the Australian National University. His distinguished academic career includes fellowships at MIT, the Indian Institute of Advanced Studies and the ANU; and service as Professor of Asian and Comparative History at the University of Michigan, and as President of the California Institute of Integral Studies, San Francisco. He brings a wealth of experience in cultures around the world, including study with Native American, Tibetan Buddhist, Hindu, Balinese, Maori and Aboriginal Australian teachers. His most recent book is *Other Ways of Knowing: Recharting Our Future with Ageless Wisdom* (Inner Traditions 1997).

Catherine E. Camden-Pratt

Catherine lectures in social ecology in the School of Education at the University of Western Sydney. Her 2003 PhD was published in 2006 as *Out of the Shadows: Daughters Growing up with a 'Mad' Mother* (Finch Publishing). She has published across creative arts, social ecology, research and pedagogy. Her most recent academic publications include chapters in *Pedagogical Encounters* (Peter Lang Publishing 2009). Catherine has exhibited her artwork, opened local art exhibitions, presented on social ecology and art making, and written and performed in plays based on research data. In 2010 Catherine received a national university teacher's award for teaching: The Australian Learning and Teaching Council's Citation for Teaching Excellence. This award was on the basis of her teaching which 'foregrounds critical creativity and establishes safe spaces for experimentation, using creative learning approaches which transform students' understanding of themselves as agents of change' (ALTC 2010).

John Cameron

Dr John Cameron worked as a geologist and green economist before spending fifteen years as a Senior Lecturer in Social Ecology at the University of Western Sydney, teaching courses on Sense of Place and coordinating the research programs. In 2005 he and his partner Vicki King moved to Bruny Island in Tasmania, where they have undertaken a land regeneration project on their fifty-five acres at 'Blackstone'. He is an Honorary Associate of the School of Philosophy at the University of Tasmania and he co-founded the Bruny Island Environment Network. John was editor of the book *Changing Places: Re-Imagining Australia* (Longueville 2003), which included contributions by 25 of Australia's leading sense-of-place researchers. His recent essays on his experiences at 'Blackstone' have been published in *Environmental and Architectural Phenomenology*.

Bruce Fell

Bruce Graham Fell is a social ecologist, academic and author. He lives in a rural valley before the junction of two creeks boarded by farmland and re-growth forest in Central Western New South Wales. Bruce writes about rakali swimming, wallaby weaving and people trapped by the World Problematique. Bruce has written and directed film, television, video and online productions. He's the author of *Television & Climate Change: The Season Finale* (www.brucefell.com). Bruce's education and community interests are directed towards questions concerning ecology, media and wellbeing. Dr Fell lectures in visual literacy scriptwriting and movie production at Charles Sturt University, Australia.

Graeme Frauenfelder

Graeme Frauenfelder is a graduate of social ecology at the University of Western Sydney where he uses experiential learning practices to teach creativity, transformation, diversity and wellbeing. His involvement with projects in Asia, Africa and the South Pacific focuses on enhancing the wellbeing and quality of life of individuals, communities and organizations by using creativity, cultural enrichment and transformative kindness. His university research in social ecology included working with Zambian professionals to empower youth, community leaders and teachers in their villages. Favorite pursuits include being an inspirational speaker and entertaining storyteller, and a playful clown at community festivals.

Susanne Gannon

Susanne Gannon is passionate about writing in and out of academia. She lives in the Blue Mountains west of Sydney, Australia, and is an Associate Professor in the School of Education at the University of Western Sydney. Susanne was a high school teacher prior to taking up an academic position. She is a co-author of *Place, Pedagogy, Change* (Sense Publishers 2011), *Deleuze and Collaborative Writing* (Peter Lang 2011), *Pedagogical Encounters* (Peter Lang 2009), *Charged with Meaning: Reviewing English* (Wakefield Press 2009) and *Doing Collective Biography* (Open University Press/ McGraw Hill 2006); and has contributed sole and co-authored chapters to the *Sage Handbook of Feminist Research: Theory and Praxis* (Sage 2007 and 2011), *Theory and Methods in Social Research* (Sage 2005, 2011), *Writing Qualitative Research on Practice* (Sense Publishers 2009) and *Poetic Inquiry: Vibrant Voices in the Social Sciences* (Sense Publishers 2009).

Robin Grille

Robin Grille is a father, psychologist in private practice and a parenting educator. His articles on parenting and child development have been widely published and translated in Australia and overseas. Robin's first book, *Parenting for a Peaceful World* (Longueville Media 2005), has received international acclaim and led to speaking engagements around Australia, USA and New Zealand. His second book, *Heart to Heart Parenting* (2008), is published by ABC Books. Robin's work is animated by his belief that humanity's future is largely dependent on the way we collectively relate to our children. http://www.our-emotional-health.com

Christy Hartlage

Christy Hartlage is a mother of two, an educator and a food lover. She has a strong belief in the ritual, celebratory power of food. Christy has pursued her interest in food, activism and cultural change through study in the USA, New Zealand and Australia. Whilst in New Zealand Christy grew herbal medicines and played a role in local and national environmental movements, most particularly the Royal Commission of Inquiry into Genetic Modification. Her current preoccupation with parenting has led her to an interest in constructing opportunities for supported care for mothers and new babies. Christy has written for a variety of popular and academic media, most recently a chapter in *Rituals Aotearoa* (in press).

Stuart B. Hill

Stuart is Foundation Chair of Social Ecology at the University of Western Sydney (now retired). His PhD was one of the first whole ecosystem studies to examine community and energy relationships and he received awards for Best PhD Thesis and Best PhD Student. He has published over 350 papers and reports. His books include *Ecological Pioneers* (with Martin Mulligan; Cambridge 2001) and *Learning for Sustainable Living* (with Werner Sattmann-Frese; Lulu 2008). He has worked in development projects across the world. His work in the Seychelles to make a coralline island self-sufficient in food and energy is particularly significant. His background in chemical engineering, ecology, soil biology, entomology, agriculture, psychotherapy, education, policy development and international development, and his experience of working with transformative change, has enabled him to be an effective facilitator in complex situations that demand collaboration across difference and a long-term co-evolutionary approach to situation improvement.

Sally Mackinnon

Sally MacKinnon has been involved in the environment and sustainability movements for over 20 years as an educator, communicator, storyteller and community volunteer. Sally contributed to the establishment of the Ethos Foundation in 2005, and her work focuses on program design and facilitation; research and writing; and participatory community engagement particularly in the areas of Local Living Economy, and community resilience and prosperity. Sally's first book, *Expanding Green Strategies: Creating Change Through Negotiation*, was published in late 2009, and in 2010 her poetry formed part of the group art exhibition 'My Black Heart' at the Scenic Rim Regional Gallery.

Kathryn McCabe

Kathryn McCabe is a senior facilitator and national program coordinator with OzGREEN. Kathryn works with businesses, schools, community groups and indigenous communities. She has lectured on sustainability education at the University of Western Sydney and presents at conferences on personal wellbeing, systems thinking and change. Kathryn applies a multidisciplinary approach of social ecology, science, applied physics, drama, therapy and education to her work for personal and cultural transformations.

John (Jack) P. Miller

Jack has been working in the field of holistic education for over 30 years. He is author/editor of more than a dozen books on holistic learning and contemplative practices in education, which include *Education and the Soul* (State University of New York Press 2000), *The Holistic Curriculum* (OISE Press 1996) and *Educating for Wisdom and Compassion* (Corwin Press 2005). The *Holistic Curriculum* has provided the framework for the curriculum at the Whole Child School in Toronto. Jack has worked with holistic educators in Japan and Korea and has been visiting professor at two universities in Japan. He recently was one of twenty-four educators invited to Bhutan to help that country develop their educational system so that it supports the country's goal of Gross National Happiness. He teaches courses on holistic education and spirituality education for graduate students and students in Initial Teacher Education at the Ontario Institute for Studies in Education at the University of Toronto.

Martin Mulligan

Martin is the Director of the Globalism Research Centre at the Royal Melbourne Institute of Technology University in Melbourne where he specializes in strategies for the sustainability of local communities in the context of global change. From 1993 to 2003 he was a lecturer in the social ecology program at the University of Western Sydney where he taught subjects related to ecological thinking and environmental sociology. During this time he worked with Stuart Hill to produce the book *Ecological Pioneers: A Social History of Australian Ecological Thought and Action* (Cambridge 2001). He also worked with William Adams at Cambridge University to produce an edited volume called *Decolonizing Nature: Strategies for Conservation in a Post-Colonial Era* (Earthscan 2003).

Bernie Neville

Bernie is Adjunct Professor of Education at La Trobe University. He holds an MA in Classics from Adelaide and a PhD in Education from La Trobe. He has been involved in the pre-service and in-service education of teachers since 1972. He has researched and written on the interpersonal aspects of teaching and learning and the application of counseling theory to the process. His particular interests in the area of classroom processes are reflected in the title of his book: *Educating Psyche: Emotion, Imagination and the Unconscious in Learning* (HarperCollins 1989). He has consulted extensively with business and educational institutions on communication within organizations and strategies for organizational change. His particular interest in archetypal psychology as a framework for exploring and analyzing organizational culture is reflected in the title of his book: *Olympus Inc.: Intervening for Cultural Change in Organizations* (Flat Chat Press, 2008).

Thomas William Nielsen

Thomas William Nielsen is an assistant professor at the University of Canberra, Australia. A member of the National Values Education Project Advisory Committee, he has served in several of the Australian Government values education projects. He is program leader of the Imagination and Education Research Group, University of Canberra branch, and has received several teaching awards, including the 2008 Australian Learning and Teaching Council Citation for Outstanding Contributions to Student Learning. Dr Nielsen advocates a 'Curriculum of Giving', his research showing that giving and service to others creates unparalleled wellbeing and resilience in students – something much needed in a Western world with rising depression and suicide rates.

Edmund O'Sullivan

Edmund O'Sullivan is a Professor of Education at the Ontario Institute for Studies in Education at the University of Toronto. He is Director of the Transformative Learning Centre that does both research and graduate programs on ecological issues that emphasizes a global-planetary vision combining ecological literacy, social justice and human rights concerns, diversity education that deals with issues of race, gender, class, sexual orientation and ableism. He is the author of eight books and has written over a hundred articles, chapters in books and refereed journals. His latest books are *Critical Psychology and Critical Pedagogy* (University of Toronto 1990) and *Transformative Learning: Building Educational Vision for the 21st Century* (Zed Books 1999).

David Russell

David is a psychologist and psychodynamic psychotherapist in private practice (East Sydney). He also holds the position of Associate Professor (Adjunct) in the School of Psychology at the University of Western Sydney. He joined The Hawkesbury Agricultural College as a lecture in organizational psychology in 1978 and taught in an adult education program, which, after a few years evolved into a set of programs in social ecology. In the year 2000, at what is now the University of Western Sydney, a group of interested faculty established a master's degree in Analytical Psychology (a course work program based on the works of Carl Jung and the post-Jungians). He is currently president of the Sydney Jung Society.

John Seed

John is founder of the Rainforest Information Centre. Since 1979 he has worked for the protection of rainforests worldwide for which he was awarded an Order of Australia Medal in 1995. He has created numerous projects protecting rainforests in South America, Asia and the Pacific through providing benign and sustainable development projects for their indigenous inhabitants tied to the protection of their forests. He has written and lectured extensively on deep ecology and co-authored *Thinking Like a Mountain: Towards a Council of All Beings* (New Society Publishers 1988). For over 25 years he has lectured on eco-philosophy and conducted experiential deep ecology workshops around the world.

Ben-Zion Weiss

Dr Ben-Zion Weiss is a community educator in social ecology, meditation, yoga, drama, English for Speakers of Other Languages, cross-cultural conflict and non-violence training. He lectures and tutors at the University of Western Sydney, consults for the New South Wales Department of Education and Training Multicultural Programs in Cooling Conflicts and other intercultural programs. His PhD research is on anti-racism drama education and an ecology of culture. He presents at conferences; facilitates workshops for youth workers, teachers and community workers; leads Dances of Universal Peace; and facilitates dialogues in spiritual ecology. He has completed a draft of a book based on his thesis and has written chapters and papers for other publications.

James Whelan

James lives in Newcastle, New South Wales in the Worimi nation. His commitment to community action for social and environmental justice has drawn him to work in the community sector and in research and higher education. James has worked with non-government, community and environment groups. James has worked with several Australian universities. He was Theme Leader for the Coastal CRC's Citizen Science research program, has published on participatory democracy and social movements, and has spoken at international conferences. James' community and academic worlds merge in his work as director of the Change Agency, which provides education, training, facilitation and research support for social change groups in Australia and the Pacific.

Peter Willis

Peter Willis is a Senior Lecturer in adult learning and education at the University of South Australia. He pioneered phenomenological approaches in arts-based research in his book *Inviting Learning: An Exhibition of Risk and Enrichment in Adult Education Practice* (NIACE 2002). His main research areas concern transformative, 'second chance' and 'resistant' learning among adults, the power of the imagination in learning, and relationships between religion, spirituality and civil society. Recent edited publications include *Pedagogies of the Imagination* (Springer 2008) (with Leonard), *Towards Re-Enchantment: Education, Imagination and the Getting of Wisdom* (Post Pressed 2005) (with Heywood, McCann and Neville*)* and *Wisdom, Spirituality and the Aesthetic* (Post Pressed 2009) (with Leonard, Hodge and Morrison).

David Wright

David is co-ordinator of Social Ecology programs at the University of Western Sydney. He is a past Head of Performance in the School of Contemporary Arts and Academic Advisor to the Department of Maori Performing Arts in Te Wananga o Aotearoa (New Zealand). David has a background in writing for performance. He has published work in a variety of styles, from creative fiction to eco-philosophy. This work has appeared in books and journals in the fields of literary fiction, applied drama, drama education, ecosophy, sense of place, eco-politics, reflective practice, imagination, cross-cultural arts practice, and higher education.

Ainslie Yardley

Ainslie is a researcher and associate member of Social Justice and Social Change Research at the University of Western Sydney. She is a novelist, theatre artist and multi-media essayist. Her work in community has included youth theatre productions and projects with refugee claimants from many areas of conflict throughout the world. She has worked with the Australian AIDS Memorial QUILT Project, the Bosnian Community Choir in Brisbane and on multimedia projects in the disability and mental health sectors. Ainslie has lectured in cultural ecology and production management. Her research practice and academic publications incorporate new methodologies and multidisciplinary approaches including embodied creativity and narrative theory.

ACKNOWLEDGEMENTS

We would like to acknowledge the inspiration and support of the many students who have studied Social Ecology units and Social Ecology programs at the University of Western Sydney (UWS) since 1987, as well as the staff, academic and non-academic, who have contributed to the continuation of this work. In particular we would like to thank Kathy Adam-Cross for her efforts over the years.

We would like to thank the School of Education at UWS (and previously the School of Social Ecology and Lifelong Learning), under the leadership of Associate Professor Steve Wilson. The school has valued the Social Ecology approach and appreciated its contribution to education. For that, and the financial support of the school for this book, we are extremely grateful.

The College of Arts at UWS has also been crucial in maintaining the Social Ecology approach. We appreciate that support.

We would like to thank Bruce Fell for his help with editing and formatting and Martin Large for his enthusiastic embrace of this project. We would also like to thank the editors and staff of Hawthorn Press for their timely and efficient efforts. Last but not least, we would like to thank our families and friends for their support in this endeavour.

INTRODUCTION
The emerging field of Social Ecology

David Wright and Stuart B. Hill[1]

We cannot know the future, but we can dare to imagine. Let us compare two contrasting scenarios. It is the year 3000, and the turn of the century is being celebrated.

In the first scenario, which confirms our worst fears, it is a severely limited event, in every sense. Only a small area of the Earth is habitable by humans, who are now a minor species on the planet, surviving much as some of the endangered species – such as the orang-utan and gorilla – are today. The survivors did eventually learn how to live sustainably, but it was too late; and 'survival of the fittest' inevitably eliminated most members of our species, together with most other species that shared our environmental requirements. It is a sad sight, but they are, nevertheless, celebrating their survival, while mourning their past and maintaining hope for the future.

In the second scenario celebrations are taking place in relatively small, largely locally self-reliant communities across the planet. These mutually supportive societies are markedly different from our own. Like the survivors in the first scenario, they are the products of intense psychosocial evolution; but the difference is that they embraced the necessary changes much earlier than did those in the first scenario. Despite the apparent 'good life' being lived by the privileged at the turn of the previous century, they recognised that this was ethically unjust and unsustainable. Perhaps most profoundly they realised that their obsession with growth in production and consumption, and neglect of system maintenance – at every level, from person to planet – was already resulting in significant degradation and system breakdown; and, if allowed to continue, that this would result in the extinction of their species. So, they set about changing everything: from their personal lifestyles to their political and economic systems, and the nature of their relationships with one another and the environment.

The details of the changes involved will, we hope, one day be written. What we can confidently say now is that this would have involved profound changes in their values; and the development and adoption of frameworks for understanding, designing, planning, relating, decision making and acting that are supportive of the well-being of all, and of all life-enabling processes. Because these processes are primarily ecological, and change involves psychological and psychosocial transformation, these are the areas where their learning and development would have been most intense.

Evidence of such thinking can be found in all areas of endeavour; and it is interesting to us that a significant number of these pioneer thinkers, who advocated applying ecological understanding to the design and management of human systems, used the term 'social ecology' (SE) to label their approach. This is the framework and approach that we are advocating and that is being explored in this collection of essays.

The pioneers who used this term included the architect and town planner Erwin Gutkind (1953), evolutionary biologist Sir Julian Huxley (1962 talk, published 1964), ecoanarchist and ecolibertarian Murray Bookchin (1964; at that time he wrote under the pen name Lewis Herber; in the mid-1990s Bookchin abandoned anarchism and proposed 'communalism' as his approach), social scientist Mattei Dogan (who in 1970 established and chaired the International Sociological Association 'Research Committee on Social Ecology'; Dogan and Stein, 1974), psychologists Fred Emery and Eric Trist (1973; this was while they were at the Tavistock Institute for Social Research in London), and behavioural scientist Martin Large (1981), who, together with Bookchin, influenced our use of the term at Hawkesbury.

John Clark (1997) has noted that since the late 1800s the ground was being prepared for the development of social ecology by those who were reflecting on the relationships between human societies and nature. Most important among these were French geographer Elisée Reclus (1830–1905), Scottish botanist and social thinker Patrick Geddes (1854–1932), his student American historian and social theorist Lewis Mumford (1895–1992), communitarian philosopher Martin Buber (1878–1965), and anarchist geographer Peter Kropotkin (1842–1921), who championed mutual aid, political and economic decentralisation, human-scaled production, and communitarian values; and who was a major influence on the work of Murray Bookchin.

There were also many important pioneers who were endeavouring to apply ecological understanding to a diverse range of fields. These included particularly sociologists Robert Park, Ernst Burgess and their colleagues at the Chicago School of Sociology (e.g. Park and Burgess 1921), which was sometimes referred to as the 'Ecology School'. Some of the other pioneers are referred to in the following chapters.

At least of equal importance to the development of social ecology thinking have been the many other pioneers who contributed to its foundations. Of particular importance was the development of an 'ecological epistemology' by Gregory Bateson (1972) in his book *Steps to an Ecology of Mind*. Bateson, along with J.J. Gibson and his book *The Ecological Approach to Visual Perception* (1979, republished 1986), drew attention away from an objective focus upon entities to an examination of the subject's relationship to the object, and in doing so were early contributors to what O'Sullivan (1999) called a reconstructive postmodern vision.

Arguably, the most influential of social ecology theorists has been Murray Bookchin. Bookchin was a prolific writer and organiser, who viewed social ecology as a political action as much as a form of understanding. His legacy lives on in the Vermont-based Institute of Social Ecology (which he founded with Daniel Chodorkoff in 1974; it was incorporated in 1981), his own writing and the life and writing of many who have been influenced by him. Bookchin settled on the term 'social ecology' as a response to the failure to deliver egalitarian social systems in a rapidly industrialising USA (associated also with a critique of socialist models on offer), and early intimations of the social and ecological consequences of that industrialisation. He writes,

> When I first began to use the rarely employed term 'social ecology' ... I emphasized that the *idea* of dominating nature has its origins in the very *real* domination of human by human – that is, in hierarchy. These status groups, I insisted could *continue to exist even if economic classes were abolished.*
>
> Secondly, hierarchy had to be abolished by *institutional* changes that were no less profound and far reaching than those needed to abolish classes. This placed 'ecology' on an entirely new level of inquiry and praxis ... Social ecology was concerned with the most intimate relations between human beings and the organic world around them. Social ecology, in effect, gave ecology a sharp revolutionary and political edge. In other words, we were obliged to seek changes not only in the objective realm of economic relations but also in the subjective realm of cultural, ethical, aesthetic, personal, and psychological areas of inquiry.
>
> (Bookchin 2002)

This identification of social ecology as an inquiry into subjectivity and relationships is in accord with the approach of Emery and Trist who published *Towards a Social Ecology* in 1973. They say they were led to social ecology by 'our concern with what was happening to organisations, considered as open socio-technical systems, as they encountered greater complexity and a faster change-rate' (1973: xii). This required a 'more thorough examination than we have made so far of environmental relations and a consideration of the character of environment itself' (ibid.: xii). Thus, environment is understood through social relationships and knowledge systems, and any change in relation to the environment is dependent on changes in social relationships and social knowledge systems.

Social ecology at the University of Western Sydney

'Social Ecology' in Australia had its origins at Hawkesbury Agricultural College (HAC) (later the University of Western Sydney: Hawkesbury; and later again the University of Western Sydney, Richmond campus). HAC was an elite agricultural college, on the north-western fringe of Sydney. It opened in 1891 and was, for many years, a conservative, male-dominated, finishing school for young farmers. Its education took the form of inculcation into the agricultural practices and social understandings of rural Australia. The urbanisation of Australia and the systematic decline in secure farm incomes, leading to a growing disinterest in careers on the land, contributed to the decline of HAC, and thence its easy absorption into the multi-campus University of Western Sydney (UWS), an institution with no historical association with agriculture, other than the Hawkesbury programs.

Within HAC, however, there were some uniquely interesting combinations of staff. Many of these remained into the early years of University of Western Sydney: Hawkesbury. Central within this was an interest in 'systems agriculture', and a funded Chair held by Richard Bawden. Bawden's leadership of the systems agriculture group, influenced by his readings of Checkland (1981) and the 'Open Systems' group at the Open University UK, laid the ground for the application of a systemic approach to learning and research. This 'Hawkesbury approach' grew out of awareness that different forms of learning are acquired as a consequence of different systemic relationships. Through this perspective, learning became much more than formally acquired knowledge. In keeping with the assumptions of contemporary andragogy, learning was regarded as self-directed, experiential, relevant and applied (Knowles 1984; Brookfield 1995): a process, rather than a content-based approach, that builds on the specific needs of individuals and communities. Accordingly, Bawden and Packham (1991) claimed a 'brand new and controversial research tradition where the emphasis is not on enquiry into systems as real entities, but as figments of the imagination of people, which help them think about real issues'.

Considerations on the personal and community relationships that sustain agriculture led to an initial postgraduate degree in Social Communication (1982), under the leadership of Graham Bird. This attracted a wide range of students: far beyond the agricultural students generally drawn to HAC. The name change to 'Social Ecology' occurred in 1987, after UWS Hawkesbury lecturer John Field had returned from a meeting with Martin Large in the UK. Large had been inspired to use this term through exposure to the work of Emery and Trist at the Tavistock Institute. Some of the staff were also familiar with the earlier use of the term by Murray Bookchin.

The methodology and structure of all courses taught in Social Ecology promoted personal understanding, which was applied to locales, practices and fields of knowledge with which the learner was directly concerned. It emphasised the centrality of relationships, and the importance of considered reflection in the construction of sustainable knowledge systems. It encouraged learning through participation and promoted inquiry through participatory action research (Reason and Bradbury 2001).

Significant early – often informal and unacknowledged – leadership was provided by women members of staff, in particular Marilyn McCutcheon, Chris Winneke and Judy Pinn. Through a focus on feminist epistemologies, experiential and process-based perspectives on learning, they contributed to the moulding of the personalised approach that made it possible for Social Ecology to be, for many years, the pre-eminent site of research training in UWS.

This is not to suggest that the Social Ecology staff group was a unified and uniquely focused one. Not only did (and do) core interests differ, but also personal and social politics contributed to what was sometimes a disrupted and disruptive learning space. At various times, sometimes in association with one of a series of all too frequent university restructures, the staff group was fractured and some left, sometimes feeling bitterly undervalued. Social Ecology has not been an easy site to inhabit.

What is social ecology?

In 1994 David Russell responded to the all too frequent question, 'what is Social Ecology', with the following.

> Social ecology is … a way of integrating the practice of science, the use of technology, and the expression of human values. It draws from any 'body of knowledge' in its pursuit of designing activities that result in self-respecting, sensitive and social behaviours, which show an awareness of social and ecological responsibilities.
>
> (Russell 1994: 148)

Stuart Hill, Foundation Chair of Social Ecology, in the opening chapter of this book provides what he calls a 'very personal account of social ecology'. He describes it as 'like finding home', partly enabled by 'our version of social ecology's integration of the personal, social, environmental and "spiritual/unknown" in most of its teaching and research', and this is reflected in the definition he provides in his chapter.

> I was also attracted by its emphasis on experiential learning, participatory action research and other qualitative methodologies, its recognition of the importance of context, and its acknowledgment of diverse ways of knowing (including women's and Aboriginal ways), the importance of diversity and of learning to collaborate across difference, of working for equity and social justice, particularly in relation to issues of power, gender and race, and of learning how to work with and design complex mutualistic systems, recognising chaos as an important precondition for creativity, development and co-evolution, and not something to be quickly controlled and simplified.

It is worth noting that it is the activity of social ecology, a way of imagining, integrating and designing, rather than any academic field or sub-field that both Russell and Hill prioritise here.

In 1999, another staff member, Brendon Stewart, did try to identify Social Ecology as an academic domain. He positioned it, reflecting his interest in Jungian/archetypal psychology, as integrating 'a "sense of place", home making, "imagination in action", community and organisational theory, the Gaia hypothesis (anima mundi), contemporary systems theory and a biology that favours symbiosis as the coherent and organising function of life' (1999: 4). At first glance this is a disparate bundle. Common ground can be found, however, in process, and the process is overwhelmingly situated in imagination, interpretation and representation. Metaphor rather than fact is to the fore: biology and culture interconnect through story, feelings are embraced and mystery is welcome.

With the election of a conservative government in Australia in 1996, universities were subjected to increasing ideological and budgetary constraints; and holistic and transdisciplinary areas such as social ecology were predictably marginalised. Reflecting on this time, Newfield (2008: 15) observed that 'the university's cultural missions have declined at the same time as leaders in politics, economics and the media have lost much of their capacity to understand the world in non-economic terms.' A major outcome for our group was that in 1998, in response to a requirement to amalgamate with other compatible units, we joined with the School of Lifelong Learning and Educational Change to form the new School of Social Ecology and Lifelong Learning; and after a further forced amalgamation in 2005 we became part of the much larger School of Education.

Although this has brought new challenges – as a small unit within a larger school – it has also opened up new opportunities. Postgraduate students in our Master of Education (Social Ecology) degree now share foundational studies with colleagues studying Educational Leadership and Special Education. They undertake units in 'Transformative Learning', 'Transformative Leadership' and 'Researching Practice', as well as 'Applied Imagination', 'Ecopsychology and Cultural Change', 'Environmental Education and Advocacy' and 'Researching Social Ecology'. Emerging synergies amongst students and across courses have opened up new pathways to more effectively enable meaningful personal, professional, social and environmental change.

At the undergraduate level, our previous degrees in Social Ecology, which at their peak had only 40 students, have been replaced by three Social Ecology units that each year are taken by hundreds of students as part of their Education Studies Major. Through their exposure to 'Learning and Creativity', 'Education and Transformation' and 'Education for Sustainability', thousands of future school-teachers have been able to actively engage with a diverse range of concepts and processes firmly located in Social Ecology. As these students progress in their careers in education and begin to develop the understandings and skills required for making a positive difference in the lives of young people, their learning in Social Ecology will, we are confident, be invaluable. It will help them to play a pivotal role as creative, reflective and self-aware educators in enabling their students to construct more sustainable, equitable, peaceful and meaningful futures.

This book is particularly relevant for those undergraduate and postgraduate students, but also for the much broader community of people seeking more ecological and humane ways to live and relate to one another and the environment.

Suggestions for using this book

The book is a response to, rather than an attempt to define the practice of, 'social ecology'. Those invited to contribute to the collection are just some of the many who have influenced and been influenced by the teaching of Social Ecology at the University of Western Sydney over the last thirty years. Full-time and part-time staff, guest teachers, authors of key texts and graduates have all contributed to this richly varied resource. Although their origins are important, it is their subject matter that connects them and marks the value of this collection.

The collection is divided into four sections, each presenting the subject matter from a different perspective. Because of the holistic and interrelated nature of the subject, the collection can be read and enjoyed in whatever order is relevant to the reader. Both the book and the subject matter encourage an eclectic, intuitive and wandering engagement. In all chapters the personal is constantly in negotiation, crisis emerges through knowing rather than ignorance, and amelioration is a consequence of attitude and reflection in relation to action. Issues of creativity, transformation and sustainability form the spine, and the future teases with learning.

The opening section, 'The Big Picture', comprises a series of articles in which worldviews are delivered, through a social-ecological perspective. In Chapter 1, current Adjunct Professor Stuart Hill's 'Social ecology: An Australian perspective' is a personal account of his experience of the philosophy and practice of Social Ecology at the University of Western Sydney, since his appointment as its Foundation Chair in Social Ecology in 1995. He describes his discovery of Social Ecology as particularly satisfying after 'having had to settle for so much less for so long'; and he identifies with its mission 'to achieve sustainability and benign change'. Central to this is the 'need to pay much more attention to neglected and blocked expressions of humanity' and the 'search for new life-promoting myths'. Similar social narratives feature also in the contribution by Edmund O'Sullivan, one of the founders of Toronto's Transformative Learning Centre. O'Sullivan's systemic analysis (Chapter 2) calls for a creative understanding of change, with respect for the 'the universe process, the earth process, the life process, and the human process within the possibilities of the historical moment'. It is his application of universe processes to education that marks O'Sullivan's contribution, both within this collection and beyond. Adult educator, Peter Willis (Chapter 3), seeks *mythos* within story. He argues that creative construction gives rise to the sort of transformative pedagogy central to the need for new ways of knowing. To be transformative, such pedagogy requires clear structure, inspiring artistry and effective delivery. He argues that story is a place in which the ideals and practices of social ecology can be imagined, and with this in mind he tells his own story of a 'place writers' workshop in southern Tasmania. By contrast, Richard Bawden in Chapter 4 looks at epistemic conflict: clashes within and between ways of knowing. His big picture, like all contributors to this section, argues the need for new attitudes and assumptions, while pondering also the ways in which old and new worldviews encounter, interact and come into conflict. Bawden argues that the social ecology perspective is one that acknowledges the responsibility inherent in knowledge. He does this in conversation with the ecological perspective and modernism's conflict with pre-modern epistemologies in indigenous Australia.

Bernie Neville (Chapter 5), archetypal theorist and teacher educator, positions his contribution in the midst of sharp social transformation, under pressure of impending ecological crisis. While recognising the failure of social systems, particularly the education system, to respond to this, Neville argues – paradoxically – for the importance of living with complexity, chaos and ambiguity. 'The fate of the planet', Neville considers, 'will not be determined by the brilliance of our technology, but by the genuineness of our dialogue and the strength of our connection to all life.' Educators have a crucial role in this.

Although the final chapter (Chapter 6) in the opening section by composer, conductor and arts educator Barry Bignell is different in tone and content to preceding ones, it is included here because of the subject matter it addresses. Bignell challenges us 'to envisage ourselves as more than we are', to be 'conscious of our humanity in all things'. Arguing this, he draws attention to 'imaginative re-creation', especially to the experience of the 'spoken word'. In doing so he asks us to consider the manner of our communion rather than the logic of it. He argues that in our naming – our languaging – we create consciousness. 'It is to the poet in the child that we must attend, because the quality of the sound-experience refines the power of observation.' The way we speak betrays the way we think; the critic needs only listen to hear us betray ourselves.

The second section of the book, titled 'The Social in Ecology', brings issues of creativity, community, sustainability, place and story into direct conversation with ecological relationships.

Ainslie Yardley (Chapter 7) identifies creativity as a physical relationship with self and environment; as a country with borders, laws and conditions of entry and exit. In doing so she maps a domain of inquiry in constant negotiation with the context. This embodied relational encounter is considered in many of the chapters that follow. Sally MacKinnon's tale of her practical and metaphoric transitions between community gardener and community activist is of this kind (Chapter 8). MacKinnon writes of 'communities as gardens – as living, evolving, self-organising organisms'. She writes about the intense 'political' experience involved in community building and gardening as a way of alleviating the oppressive spirit let loose.

In their contribution, recent Social Ecology graduates Jasmin Ball and Kathryn McCabe (Chapter 9) argue for the need to engage critically and actively with sustainability; and they present this as personal dilemma, not a problem of or for others. They advocate that we 'feel' our way into change through an appreciation of 'mutually supportive relationships'. Using examples from their activist work, they deeply ponder the problems involved in taking effective action.

In Chapter 10 John Cameron, long time Social Ecology staff member and major voice in 'sense of place' discourse, writes of his pathway to an appreciation of place and the ramifications of a deep and abiding relationship with it. Cameron describes his approach as emerging through Social Ecology's focus on experiential learning and reflection. By positioning these in relation to repeated encounters with specific locations, Cameron seeks to bring to the fore conversations between students about 'their place' and their learning. This is a process that Cameron himself has lived. It has contributed to his retirement from UWS and the reconstruction of his present low-impact lifestyle, and the regeneration of fifty-five acres of degraded land in a remote location on Bruny Island, Tasmania.

Martin Mulligan, like Cameron, is also a former staff member of Social Ecology. In Chapter 11 Mulligan writes of his gradual engagement, post-UWS, with local and global communities. His focus is how communities hold knowledge. He identifies conflict in the relationship between the knowledge systems of global organisations and local communities, and argues that resolving this is central to the development of effective climate change politics. Bruce Fell, Social Ecology graduate and documentary filmmaker, approaches the 'social in ecology' through reference to memory systems and technology (Chapter 12). 'Neuroscience, in combination with cognitive archaeology, informs us that memory is located in two places: internally and externally. This chapter is about this relationship.' By positioning memory in powerful technology, outside the central nervous system, Fell questions the ways in which civilisations can upgrade and contemporise information that is crucial to human well-being and ecological sustainability.

In the following chapter (Chapter 13), the final one in this section, archetypal psychologist and former Social Ecology staff member David Russell echoes aspects of Fell's analysis. Although not addressing technology, Russell examines the relationships between the collective imagination, the construction of mythologies and contemporary life-issues, such as climate change. Russell regards the challenge of 'engaging our imagination in the task of wrestling with real world problems' as first and foremost involving psychological work. Long-standing images and metaphors – such as the earth as passive, nurturing and supportive – can impede this. Russell highlights the need for emotional desire to drive our imagination. 'Desire moves', Russell asserts, 'things change … and we have reason for hope'.

All chapters in this section identify creativity as a means of engagement. All position the experience of relationship deep within the knowledge systems that determine social-ecological understanding. The application of these to learning systems is central in the section that follows.

The third section of the book, titled 'Education and Transformation', opens with an essay (Chapter 14) by Canadian holistic educator John P. (Jack) Miller on Henry David Thoreau. Miller describes Thoreau as both an environmentalist and an educator; who provides a model for effective teaching. Shortly after concluding his tertiary studies, Thoreau established his own school; and central to the program provided were field trips. In this way, nature became source material for all disciplines. The closure of the school, following the death of his brother, triggered Thoreau's immersion in nature at Walden Pond. Education and learning was never far from his thoughts. Miller characterises Thoreau as someone who wrote 'to inspire the individual to awaken and to live the life they can imagine'. He was, Miller argues, 'one of the first environmental educators'.

Thomas Nielsen (Chapter 15) also considers education as a means for developing positive relationships. He writes about a program designed to educate the benefits of generous action: a curriculum of giving. However, Nielsen writes, 'without giving to the self, with wisdom and awareness, what the self needs, it is hard to give effectively to others'. Thus, Nielsen regards engagement, meaning and happiness as all being within the ambit of school education. He describes a variety of 'giving' initiatives, and the sites of their enactment; and he argues that it 'makes sense to view acts like gratitude, reverence, awe, prayer, etc., as ways of giving to life itself'.

Roslyn Arnold (Chapter 16) has a strong interest in empathy and learning. On the basis of her research into neuroscience, she extols the social-ecological perspective of conscious relationship: 'that ability to experience one's self as a separate being from others, but as a dependent being too'. Arnold advocates this as an invaluable quality for teachers. The capacity to tune into the needs of others, to appreciate the internal dynamics of individual class members, is that which enables transformative understandings to emerge in a classroom setting. Psychotherapist Robin Grille (Chapter 17) is also concerned with neuroscience and nurturing. He argues, 'in childhood and adolescence, the human brain is subjected to profound chemical and synaptic changes wrought through the impact of human relationships. These changes underpin the formation of individual personality and relating styles: the building blocks of any society.' Frustrated, even angered, by the inadequacy of his own schooling, Grille seeks to realise the dynamics that enable the release rather than the neglect of children's 'unique and diverse passions'.

In the next two chapters, current Social Ecology staff member David Wright (Chapter 18) and Social Ecology graduate Graeme Frauenfelder (Chapter 19) build their discussions around the social-ecological learning acquired in the practical processes of drama. Wright argues that within drama processes lie opportunities to acquire a deeply embodied appreciation of the environmental interrelatedness that constructs ecological understanding. Working through principles of cognitive biology and dramatic improvisation, Wright places value on the 'state of becoming' central within drama experience. Frauenfelder's discussion is built around play and joy. He writes of the inspiration acquired in community education work in Zambia where, with a troupe of actors, he travelled from village to village using drama to help build and strengthen community life. He writes also of his participation in community festivals in earthquake shattered China and racially riven South Africa. Play becomes for Frauenfelder an exploration and celebration of spirit, and a manifestation of social-ecological learning.

Current Social Ecology staff member Catherine Camden-Pratt (Chapter 20) comes to the heart of social ecology and its possibilities in personal becoming, through a focus on writing as practice in creative learning with/in social ecology and the consequences of this for her teaching. She acknowledges the blank page and its power with confidence in her vulnerability and uncertainty as she writes to know. Her writing demonstrates embodiment and how to write this into an academic context. The writing of the chapter itself becomes her subject matter, intermingling with the difficult questions she asks of herself, of creativity and the nature of becoming with/in social ecology. As she observes, 'This is ecological writing that acknowledges the relationships and the contexts in which the writing takes place and their influences on the writer/writing.' Creativity is, for Camden-Pratt, a negotiation between skills, capacity and the unknown. The tools she calls up are multiple, and the relationships she constructs are among the valuable legacies of the learning she communicates.

The final section, 'Ecological Stories', draws together very personal storied responses to the experience of emergence within and through ecological crisis. In Chapter 21 Christy Hartlage draws attention to the practiced rituals that enrich everyday life. Central to these rituals are our relationships with food. These, Hartlage observes, provide a commentary upon our relationships with the Earth. Here lies insight into place, into cycles of life, into production processes, and into the values that inform the depth of our daily communion. Hartlage observes: 'Understanding that our relationship with the Earth is our primary relationship: the relationship that keeps us alive, can lead us to a sensual relationship with our natural community.'

Current Social Ecology staff member Carol Birrell (Chapter 22) also offers a deeply personal tale of relationship. In this instance, it is with a particular region and its people. She invites her readers to accompany her 'deeply' into Aboriginal country; and asks how, or indeed if, it might be possible to 'think black' in Australia. She depicts this as 'moving toward' understanding; as a 'surrendering into … the land and into another way of being'. Birrell asks us to imagine her encounter: to listen for its silences. She tells of those who enabled her to develop this relationship, and about the conflicts encountered along the way, on land, sea and in dreams. She tells of being watched, and being seen here, and asserts that 'an in-depth engagement with Aboriginal culture on its own terms is required. If one desires to sit comfortably with this land, surely one needs to surrender to the land on its own terms', she writes.

James Whelan (Chapter 23) is a community activist and organiser; and in this role he positions himself in the midst of a protest march in Canberra and ponders the relationships that accompany direct action. He describes the police performing to script, and the protestors responding in kind. 'Quickly, a routine was established. In response to their script, we replied, "I will not cooperate with a government unwilling to act to prevent catastrophic climate change."' One by one, activists were led, carried or dragged away, 'their faces communicating fear, conviction, concern and solidarity'. Whelan advocates an ongoing role for direct action in a struggle informed as much by doubt in personal and community resilience as by the possibility of success. Without pondering what success entails, Whelan fears the consequences of what he calls 'the alternative'.

John Broomfield's contribution (Chapter 24) is a primer on shamanism and its contribution to understanding the breakdown in relationships between Earth and humans. From his perspective there should be a considerable amount of unknown admitted to this discussion. He observes: 'By our ancestors' measure, we have grossly exaggerated our self-importance in the intricate web of life.' In doing so, we have failed to recognise the likelihood that 'there are many more shoulders sharing this burden than we think'. In the chapter that follows, John Seed, with David Wright, also writes from a depth perspective (Chapter 25). His subject matter is the anthropocentrism that is central to the Judeo-Christian tradition. One consequence of this is, he argues, an attitude to the Earth as no more than a resource awaiting commodification. Seed identifies this as a theology that has taken root in economics and among economists. He interprets the consequences of this in the language of ritual sublimation before asking: but what are we to do? Seed advocates an activism of the spirit; as well as the confronting application of humour. Although Seed's debt to Thomas Berry is strong, his debt to his twenty-five years of work as an activist, deep in the mulch of the rainforest, is as important in his contribution to the deep ecology movement.

In the chapter that follows, Social Ecology graduate Ben-Zion Weiss writes (like Wright and Frauenfelder) of the practical application of drama, in his case, for education in anti-racism (Chapter 26). Weiss tells of his own discovery of drama as a means of constructing culture; and he argues for using this as a means for constructing an appreciative culture, able to invest in relationship. He writes of his own experience of racism, and the importance of framing it as an example of 'cross-cultural' conflict. Herein lies the opportunity to use strategies acquired in peace and conflict studies to find resolution.

In the final chapter (Chapter 27) current UWS School of Education staff member Susanne Gannon maps her own relationship to place, in poetry: her 'idiosyncratic response to the call to engage with my particular place and space, through my particular preferred medium of language'. She describes her poetry as a 'material space' in which all components of an environment co-mingle. As she walks she encounters not only a physical landscape, but also a landscape of memories and imaginings. The neighbourhood becomes a 'central protagonist' in a journey of knowledge. Gannon posits poetry as a form of inquiry: a way of knowing and articulating a depth of relationship. An evocation of social-ecological knowledge.

This is a rich collection of readings. We hope you find much that is stimulating and rewarding within. The subject matter is not new, but the times in which we live make it vital and compelling. Experience transforms authority and interrelationship becomes the key to working with the barely fathomable change we are immersed in. As you find meaning in this collection we invite you to find your own voice and your own stories, and to speak with your own communities about your learning and your understanding of social ecology, knowledge and the future we are co-creating.

Note

1 The authors appreciate suggestions prior to and during the writing of this introduction by Catherine E. Camden-Pratt and Brenda Dobia.

References

Bateson, G. (1972) *Steps to an Ecology of Mind*, Northvale, NJ: Jason Aronson; reissued in 2000, Chicago, IL: University of Chicago Press.

Bawden, R.J. and Packham, R.G. (1991) 'Improving agriculture through systemic action research', in Squires, V. and Tow, P. (eds) *The Nature and Dynamics of Dryland Farming Systems*, Oxford: Oxford University Press.

Bookchin, M. (1964, initially under the pen name Lewis Herber) 'Ecology and revolutionary thought', *New Directions in Libertarian Thought* (September); republished in 1971 in *Post Scarcity Anarchism*, Montreal: Black Rose Books, 55–82.

Bookchin, M. (2002) 'Reflections: An overview of the roots of social ecology', *Harbinger*, 3(1), http://www.social-ecology.org/2002/09/harbinger-vol-3-no-1-reflections-an-overview-of-the-roots-of-social-ecology, accessed 2 December 2010.

Brookfield, S. (1995) *Becoming a Critically Reflective Teacher*, New York: Jossey-Bass.

Checkland, P. (1981) *Systems Thinking, Systems Practice*, Chichester, UK: John Wiley & Sons.

Clark, J. (1997) 'A social ecology', *Capitalism, Nature, Socialism,* 31: 3–33.

Dogan, M. and Stein, R. (eds) (1974) *Social Ecology*, Cambridge, MA: MIT Press.

Emery, F. and Trist, E. (1973) *Towards a Social Ecology*, London: Plenum.

Gibson, J.J. (1986) *The Ecological Approach to Visual Perception*, Hillsdale, NJ: Lawrence Erlbaum Associates.

Gutkind, E.A. (1953) *Community and Environment: A Discourse on Social Ecology*, London: Watts.

Huxley, J. (1964) *Essays of a Humanist* ('Education and humanism', p132), Harmondsworth, UK: Penguin.

Large M. (1981; reissued 1996) *Social Ecology: Exploring Post-Industrial Society*, Stroud, UK: Hawthorn Press.

Knowles, M. (1984) *Andragogy in Action*, San Francisco, CA: Jossey-Bass

Newfield, C. (2008) *Unmaking the Public University*, Cambridge, MA: Harvard University Press.

O'Sullivan, E. (1999) *Transformative Learning*, London: Zed Books.

Park, R.E. and Burgess, E.W. (1921) *Introduction to the Science of Sociology*, Chicago, IL: University of Chicago Press.

Reason, P. and Bradbury, H. (2001) *Handbook of Action Research*, London: Sage Publications.

Russell, D. (1994) 'Social ecology – education and research', in Fell, L., Russell, D. and Stewart, A. (eds) *Seized by Agreement – Swamped by Understanding*, Glenbrook, NSW: Drs Fell, Russell & Associates, 147–65.

Stewart, B. (1999) 'Editorial' in *A Social Ecology Journal 1999*, Richmond, NSW: UWS Hawkesbury, p.4.

PART 1

THE BIG PICTURE

1

SOCIAL ECOLOGY

An Australian perspective[1]

Stuart B. Hill

This is a very personal account of social ecology. In this chapter I will endeavour to discuss what social ecology means to me at this moment, place it within the vast smorgasbord of frameworks for understanding and action, share some critical moments in my evolving love affair with it, and talk boldly about where I believe it can make important contributions to our future, from the individual level to that of the species, and from the local to the global.

Social ecology at the University of Western Sydney

First let me say that social ecology at the University of Western Sydney (UWS) is significantly different from the usual textbook descriptions, which invariably refer only to the writings of Murray Bookchin (e.g. Eckersley 1992; Merchant 1994: 8–9). I have not been able to find clear evidence for Bookchin's first use of the term, although it was probably in the mid-1960s; and his first major work on social ecology was *Post-scarcity Anarchism* in 1971, although he had published an earlier 'preparatory' work under the pseudonym Lewis Herber (1962). Marshall (1992: 423), however, notes that the American ecologist E.A. Gutkind (1953) was the first to refer to social ecology in a publication (it was also a term used by Sir Julian Huxley in a talk at the University of Southampton in 1962, and included in a collection of his essays, 1964), although Bookchin was the first person to develop it into a field of study with a set of principles. These, which I broadly embrace, include unity in diversity and complexity, spontaneity, complementary and mutualistic rather than hierarchical relationships, active participatory democracy and bioregionalism. Although Bookchin has written a lot about a lot of things, he is most known for his disappointments (e.g. with ecologists, Bookchin 1980; see also my reply, Hill 1980a, and observations by Smith 1998: 79) and dislikes – notably hierarchical systems, mysticism, primitivism, postmodernism and deep ecology (Bookchin 1995). At UWS, we tend to be much less judgemental in these areas. Bookchin (who died in 2006) was a passionate ecoanarchist and ecolibertarian who was eager to warn people about the dangers of most aspects of our current society, and to provide us with a critical view of our political history (Bookchin 1982). His central historical position was that domination of nature has its roots in the domination of humans by other humans, first on the basis of age and gender, and later also race and class.[2] Whatever their origins, because all of these 'dominations' have been systematically institutionalised and integrated into most cultures, an acknowledgment of our interdependent relationships with nature, and of the need for the promotion of non-hierarchical cultures, is particularly challenging. Despite the difficulties of fully understanding Bookchin's position (Watson 1996; Clark 1997; Biehl 1998: ix; see also many papers in Light 1998 for a philosophical and historical analysis of Bookchin's social ecology), he has enjoyed a significant following, particularly in the New England States where he inspired students at Goddard College in Montpellier, Vermont, at which degrees in social ecology have been offered since the 1970s.

For me, as for most of my colleagues and our students at UWS, finding social ecology was like finding home, a home that many of us had almost given up believing might really exist, having had to settle for so much less for so long. This unwillingness to settle for less and a passion to go further, particularly in our understanding and action relating to sustainability and change, is for me one of the most attractive features of social ecology at the University of Western Sydney. This is partly enabled by our version of social ecology's integration of the personal, social, environmental and 'spiritual/unknown' (discussed below) in most of its teaching and research. I was also attracted by its emphasis on experiential learning, participatory action research and other qualitative methodologies, its recognition of the importance of context, and its acknowledgment of diverse ways of knowing (including women's and Aboriginal ways), the importance of diversity and of learning to collaborate across difference, of working for equity and social justice, particularly in relation to issues of power, gender and race, and of learning how to work with and design complex mutualistic systems, recognising chaos as an important precondition for creativity, development and co-evolution, and not something to be quickly controlled and simplified.

Social ecology brings together so many poles that rarely meet: the arts and sciences; critical thinking, reflexivity, passion and intuition; rationality and spirituality; the stories of the ancients, systems theory and chaos theory; plus an extensive list of disciplines. Our social ecology is a transdisciplinary metafield that has been particularly informed by ecology, psychology and health studies, sociology and cultural studies, the creative arts, holistic sciences, appropriate technology, post-structuralism and critical theory, ecofeminism, ecopolitics, ecological economics, peace and futures studies, applied philosophy and 'spirituality' (in its broadest sense).

Minimal competencies for working as a social ecologist

To work effectively as a social ecologist one requires competencies in a number of areas that are rarely grouped together in educational programs, particularly at the tertiary level. These include certain minimal understandings in the following four areas.

- Personal: what it means to be fully alive as a member of our species and of one's communities, and as an active, responsible and creative partner in relationships (Shem and Surrey 1998), how our bodies and minds work (see especially the ecological epistemology provided by Bateson 1972, also Harries-Jones 1995), how we learn and develop, the relationships between the physical, mental, emotional and 'spiritual', between organism and environment (and sense of place), self, others and society, and a basic understanding of physiology and psychology.
- Social: including the nature of our various institutional structures, instruments and processes (politics, economics, religion, the arts, science and technology, education etc.), and our history, particularly our psychosocial history (see especially deMause 1982 and 2002 for a challenging view of this).
- Ecological: biodiversity (Dale and Hill 1996), biophysical processes, time and space, niches, roles and multifunctionality, limits and thresholds, non-linearity and cycles, mutualism and synergy, ecological succession and co-evolution, resilience, and self-regulation and maintenance.
- Processes of change: relations between personal, social and environmental change, the driving and restraining forces that are involved, and how to strengthen and add to the former, and weaken and remove the latter (Lewin 1935), how to work with 'memes' (Beck and Cowan 1996), imagination, creativity and visioning, collaborative inquiry (Heron 1996), participatory action research (Reason and Bradbury 2001), soft systems methodology (Checkland and Scholes 1990) and mutuality, or the 'we' as Shem and Surrey (1998) refer to it.

In gaining understanding and competence in each of these areas it is necessary to have opportunities to learn through personal experience and through exposure to mixed outcome case studies and diverse models.[3]

Definition

Because of the richness referred to above, it is difficult to find a definition of social ecology that is widely accepted within the UWS community, partly because the parts of it that are emphasised by each individual vary with their current interests and contexts. For me, at the moment, it is concerned particularly with 'the study and practice of personal, social and ecological sustainability and progressive change based on the critical application and integration of ecological, humanistic, community and "spiritual" values'. I am aware that all of these terms are hotly contested. However, I am choosing to use them, with some degree of discomfort, until I find better ways to describe my position.

Inclusion of the personal, social and ecological

Such a condensed definition needs some explanation. Let me say first, however, that by stating my latest provisional thinking on the values that I consider central to social ecology I am hoping to encourage others to do likewise, partly to help me to further develop my own understanding. I am making the following statements not to say that this is how it is or must be, but rather that this is how it seems to me at this moment in time. It is my current story, my collection of narratives that make some sense of my experiences as a social ecologist. Working with such embodied stories is also central to my practice as a social ecologist.

The first and, for me, most important point is the explicit inclusion of the personal, emphasising our relational self (Josselson 1996; Shem and Surrey 1998). Most comparable so-called holistic frameworks for understanding and acting in the areas of sustainability and change use as their three main categories economics, society and the environment. I believe that this privileging of economics, as being more important than all of our other social constructions – more important, for example, than politics, religion, the arts, science and technology, education, systems of values and ethics – is part of our problem. It helps to perpetuate a narrow monetary system of values and decision making; and, by doing this, it concentrates power in the hands of those with large amounts of money. A broader and more diverse base for decision making would be more compatible with, and supportive of, a participatory democracy. Also, the common neglect of the personal supports the widespread perception that our problems can only be solved by heroes (mythologised rather than real people), particularly politicians and scientists, rather than problem solving (and, more importantly, prevention) being a collaborative project that requires all of our contributions.

Money, along with our other institutional structures, instruments and processes is, I believe, better regarded as a 'tool' that needs to be designed/redesigned and used wisely to help us to implement our values. Such tools need to be subservient to and supportive of our collective broader values, and not the other way round. Taking such an approach would cause us to pay much more attention to the development and clarification of our values and to their centrality in our day-to-day discourse, decision making and action. So many of the crises reported in the news each day provide clear feedback that most of our institutional structures, instruments and processes are in urgent need of redesign. Yet there is an enormous resistance to both acknowledge this and embark on the necessary task of transformation and redesign. Rather, the usual response is to seek solutions to the symptoms of crises within a problem-solving framework.

My own vision of a preferable society, based on my present limited level of understanding, would have the following features. A right to meaningful work and access to the ingredients needed to construct healthy and creative lives would gradually replace our current view of 'labour' as a cost to be minimised and even eliminated. This might help us to recognise the current dominant attitude as just one expression of our enslavement to a manipulated, deceptively simple economic bottom line (when the absolute bottom line is bio-ecological). With a more widespread recognition of the importance of ecological limits and opportunities, the development and use of solar and appropriate technologies would be emphasized; and non-renewable resources would be conserved for higher priorities than running cars and heating houses. The need to conserve the rich biodiversity with which we share this amazing planet would be much easier to understand. This contrasts with our current oversimplified division of nature into resources to be managed and sold for profit, and pests to be eliminated with the vast chemical arsenal that we have assembled to, tragically usually non-specifically, eliminate life. What is most puzzling about this situation is that most people seem to assume that they are somehow immune to these non-specific attacks. We should expect, rather than be surprised by, the common increases in degenerative diseases, immune system breakdown and associated behavioural problems. Indeed, these should be regarded as indicators to be responded to at the causal level, rather than as new 'enemies' to be subjected to the same faulty thinking and overkill technologies that got us into this mess in the first place. Most of the new biotechnology 'solutions' are sadly being conceived within this same deceptively simple construction of nature. This time, however, the ability of naively reconstructed organisms to multiply themselves and conduct their own 'experiments' could lead to much greater crises than those associated with our naive physical and chemical experiments.

Norgaard's (1994) co-evolutionary framework for sustainable development and change similarly stresses the importance of values. It also highlights the tendency of our overemphasis on powerful institutional structures, such as global economics and the transnational corporations that they serve, together with certain powerful technologies, to colonise and compromise our values, diverse knowledge systems, and the health of our environment, our communities, our relationships and ourselves. He argues that genuine sustainable development requires a tick in each of these areas without such compromise.

My experience of working with a diverse range of populations over the past 40 years has led me to conclude that situation improvement projects are most effective and sustainable when they work in ways that integrate the personal, social, ecological and 'spiritual'. Such a framework is most supportive of participation, collaboration, personal development and creativity, responsibility and ownership, and a sense of place, purpose and meaning (Hill 2003). Thus, for me, the difference between highlighting the personal or economic is far from trivial. It has important consequences, not least of which is the imperative for developing the competencies noted above in each of these four broad areas, and gaining an understanding of the ways in which they are interrelated.

Sustainability and progressive change

I regard sustainability and progressive change very broadly. For example, at the personal level I regard most psychotherapy as being concerned with sustainability (through its involvement in recovery, rehabilitation, reconstruction and maintenance, especially of mutually beneficial relationships) and progressive change (through transformation and development). Similarly, within societies most of our institutional structures, instruments and processes are preoccupied with the often apparently contradictory forces associated with sustainability and change (Sattmann-Frese and Hill 2008).

The highlighting of sustainability and progressive change emphasises the two dominant features of all living systems; maintenance, into which most resources and energy are naturally channelled (usually well over 50 per cent), is the process that enables sustainability, and adaptation, transformation, development, succession and co-evolution are the dominant expressions of change.

The tension between sustainability and change is similar in some respects to that between the two interrelated main ways of being in the world that are essential for our well-being: 'knowing' (as a necessary prerequisite for effective and responsible action) and 'unknowing' (necessary for learning). This is best visualised as a progressive upward spiral, with knowing on one side, and learning on the other. The key is to be able to move flexibly and appropriately between these two processes, not getting stuck for too long on one or the other side of the emergent spiral. If we get stuck on 'knowing' we are in danger of becoming boring, oppressive and controlling 'know-it-alls', with well developed defences against new learning. People stuck on 'unknowing', on the other hand, often present as being apathetic, lost, searching, postponing, or as hypercritical 'unknow-it-alls'. Learning to work flexibly and spontaneously with knowing and unknowing, the rational and the mystical, science and 'spirituality', the modern and postmodern (and post-postmodern), order and chaos, goals and plans, visions and dreams, and sustainability and change is one important expression of the essential competence of being able to embrace, learn from and work with paradox, an essential social ecology competence that remains largely undeveloped in our society. Rushkoff's (1996) book *Children of Chaos* provides a paradoxical, challenging and hopeful view of how many young people, by playing with chaos, creativity and 'shadow' material, and with computers, are already intuitively preparing themselves for creating a more benign and caring future.

With respect to sustainability – the rehabilitation, conservation and maintenance of ecological, cultural and personal capital, including especially mutually beneficial relationships – it is important to recognise that whereas ecological (and, to some extent, personal) sustainability deal with absolutes, such as the air, water and nutrients for life, together with a vast range of mutualistic relationships, the requirements for social and cultural sustainability are relative and much more flexible. Because money has no comparable requirements, economic sustainability, in contrast, is dependent primarily on the wisdom of our decisions and actions (thus, money is in no way comparable to resources like air and water). Consequently, economic sustainability must serve firstly ecological sustainability, and secondly personal and sociocultural sustainability, and not vice versa. P.A. Yeomans (1958) Keyline system provides an exemplary systemic approach for working ecologically with landscapes (see also Savory and Butterfield 1999 for a systematic approach to decision making and its application, particularly to range management).

The absolute nature of ecological sustainability has important legislative, legal and regulatory implications. Thus, interventions into ecosystems (simplification, harvesting, waste disposal, release of novel chemicals and genetically modified organisms) must be regarded as 'guilty until proven innocent', with much reliance being placed on the precautionary principle (Harding and Fisher 1999; Raffensperger and Tickner 1999). Similarly, risk studies need to be conducted with reference to ecological absolutes and sociocultural values, and not based simply on economics. Currently, as part of a tendency to preserve the status quo (or extrapolations of it), most risk studies, which should be providing us with valuable feedback for necessary 'redesign', are concerned only with problem measurement and assessment – I call this 'monitoring our extinction' research – rather than with risk reduction and avoidance (see also Raffensperger 1998).

With respect to change, it is important to distinguish between 'deep' sustainable change, which usually requires fundamental redesign of the systems involved, and of our relationships with them, and 'shallow' adaptive, substitutive and compensatory change, which usually unintentionally protects and perpetuates the very structures and processes that are the sources of the problems that we are endeavouring to solve. In my work I distinguish between '**E**fficiency', '**S**ubstitution' (shallow) and '**R**edesign' (deep) approaches to change (e.g. Hill 1998). Although this **'E-S-R'** model was first developed for re-conceptualising pest control from inefficient to efficient use of pesticides, to the use of substitutes such as biological controls, to the integrated redesign and design of complex agro-ecosystems – to favour the crops and natural controls and not the pests, e.g. Hill 1990; Hill et al. 1999 – I have found it to be broadly applicable to all systems. It is important to be aware that 'efficiency' and 'substitution' strategies may serve either as stepping-stones or as barriers to the ultimately needed 'redesign' approaches.

My broad use of the ESR model, and habit of suspecting that any phenomenon detected in one part of a system might also be operational in its other parts – in generically similar yet specifically different ways – is one expression of the concept of 'holonomy' (Harman and Sahtouris 1998), and of the 'holographic paradigm', which is central to my approach to social ecology.

I have also found that when working with social change it is important to meet people where they are, acknowledge their past and present relational efforts, support their 'next small meaningful steps', and if appropriate to celebrate their progress and completions publicly to facilitate their spread. This is in contrast to the more common overemphasis on ('Olympic' scale) mega-projects, heroes, experts and heavy-handed technological and legislative interventions. Using the former approach with Quebec farmers interested in adopting more sustainable systems of farm design and management led to much higher rates of change than had been achieved elsewhere using the more conventional top-down approaches (Hill and MacRae 1992).

Another key to the effective implementation of sustainable change is to be imaginative in integrating personal (including 'spiritual'), social (including institutional) and environmental approaches, while also being aware of their (limited) substitutability. For example, the provision of a benign environment may, even in the absence of personal change initiatives or the fundamental redesign of institutions, lead to benign behaviour and health. This was achieved most dramatically in the Peckham Experiment in the UK. In this experiment, over 1,000 families, who had access to a supportive recreational centre in Peckham between 1935 and 1950, experienced no marriage breakdowns, no violence, little interest in competitive games, the widespread formation of mutually beneficial relationships and a dramatic improvement in their health and well-being (Williamson and Pearse 1980; Stallibrass 1989). Similarly, there are numerous examples of individuals in deep psychotherapy, or who have been members of a supportive peer counselling or relationship counselling group, in the absence of environmental or institutional changes, significantly transforming their ways of being and relating in the world and similarly achieving dramatic improvement in their health and well-being (Janov 1971; Mahrer 1978; Gruen 1988; Stettbacher 1991; Jackins 1992; Rowan 1993; Shem and Surrey 1998). The greatest gains are likely to be achieved, however, when mutually supportive and potentially synergistic initiatives are being taken, in integrated ways, in all of these four areas (personal, social, ecological and 'spiritual').

Conversely, it is not surprising that in a culture that emphasises growth, greed, individualism, power over, hierarchy and compensatory, stimulatory consumption (particularly through commodification and manipulative advertising), and other characteristics listed in Table 1.1, that disempowerment, relationship breakdown, apathy, irresponsibility, addiction and violence will be common. Clearly, if we are to achieve sustainability and benign change, we will need to pay much more attention to the neglected and blocked expressions of humanity listed on the right of this table.

Ecological, humanistic and community values

Because reference to these qualities has been made throughout this chapter, only certain contentious points will be highlighted here. These values need to be considered together to avoid arriving at conflicting imperatives. As indicated above, however, it is essential that our species recognises (at every level) the primacy of those ecological values (listed earlier) that are concerned with our survival, health and well-being. Because these are currently being compromised within our societies by so many political, cultural, business and personal decisions, this point cannot be overemphasised. I have previously published a more extensive list of ecological values (Hill 1980b), and have spent over 40 years endeavouring to apply them, particularly to the design and management of sustainable food systems (e.g. Hill 1998, 1999).

Enlightened humanistic values (Bookchin 1995) demand that we live up to our potential as human beings. There is currently much confusion and polarisation

Table 1.1 Dominant pressures and areas of neglect in industrialised societies

Dominant grand narrative of 'progress	*Neglected/blocked*
Production (regardless of cost)	Maintenance, caring
Growth, no limits	Sustainability, limits (resources, ecological …)
Competition	Collaboration, mutualism, synergy
Wealth	Sense of enough
Individualism	Community, mutualistic relationships
Consumerism (emphasising compensatory wants)	Conserver society (meeting basic needs)
Homogenisation, simplification	Maintenance of diversity
'Controlling' science ('understanding' science and arts as a disposable luxury)	'Understanding' science and arts
Powerful technologies (often centralised, imported, inaccessible, unrepairable)	Appropriate technologies (decentralised, locally accessible, repairable)
Market forces (manipulated demand, excessive advertising)	Values-based decisions (participatory democracy)
Economic rationalism (monetary system of values)	Meeting the greatest 'good' (social justice …)
Transglobal corporate managerialism	Regional self-reliance and responsibility
Mobile, disposable workforce (disconnected from place)	Sense of place, right to meaningful work
The *myths* that these are embedded in are *inadequate* for securing a 'good' future for most in present and future generations	We *need* to search for *new life-promoting myths* that can accommodate these characteristics: *some* can be found *within nature* (*and ecology*)

in this area. Social ecology has been accused of being overly anthropocentric by some of its critics, who have compared it with the supposedly more biocentric deep ecology perspective of Arne Naess (1989) and his followers. My version of social ecology is, however, constructively critical of both positions. Because all healthy humans naturally have a survival instinct, we are, in this respect, innately anthropocentric. To value another species above, or exactly equal to, that of one's own can often be indicative of deeper problems of psychological woundedness, transference and/or projection. For example, if as a child one's 'animal' nature was inadequately acknowledged, nurtured and integrated into one's personality, one adaptive compensatory response might be to seek alternative external ways of keeping this alive, perhaps through an excessive concern for other species, especially those with which we most commonly identify. My point here is that by raising children to value their 'animal' nature (along with their other natures) they are more likely to be proactive in valuing the richness and diversity of nature as a whole, and to be consistent in acting on this knowing in responsible ways. In contrast, compensatory preoccupations tend to be relatively temporary and the energy invested is often more cathartic than constructive. The other extreme adaptive response to such deficient child rearing might be to largely deny one's 'animal' nature and, in so doing, also the value of external nature.

A parallel argument has been applied to valuing and nurturing the feminine, as well as the masculine, in males, and the masculine, as well as the feminine, in females (Shem and Surrey 1998). Certainly we live in a world dominated by patriarchy, androcentrism, extreme anthropocentrism, technocentrism, racism, ageism, and a range of other uncaring and irresponsible prejudices. Clearly these must be addressed if we are to not disadvantage future generations and further diminish the planet's biodiversity and habitat quality. Trying to resolve these problems by fanatically focusing on a particular type of 'otherness' tends to lead to further problems, not least of which is a common lack of respect and heightened competition between those committed to different 'others'. The key, I believe, is to develop our understanding and caring for both our selves (our diverse natures) and otherness. Part of the common concern for making these equal (rather than equitable), simply by taking from one and giving to the other, may come from an assumption that there is not enough caring (or resources) to go around (another common 'lesson' from childhood). The personal task is to respect, value, support and develop mutualistic relationships with others so that their needs may be satisfied and their creativity and 'gifts' to the world expressed and received. The social task is to create contexts that are supportive of doing this, and especially of nurturing humanistic values and mutualistic relationships in children. Key child rearing and personal development references that integrate this awareness include Solter (1989), Stallibrass (1989), Josselson (1996), Shem and Surrey (1998) and Sazanna (1999).

The importance of community values follows from the above, for children need to be raised in diverse interactive communities in which they feel cared for and where they can form meaningful relationships. Although the current widespread loss of community, unlike the loss of species, is largely reversible, it is nevertheless a source of immense pain and diverse compensatory consumptive and impacting behaviours. It is also an example of an externality that is rarely considered in our obsession with short-term economic efficiency and associated economic rationalism and managerialism. A hopeful development is the growing literature on cultural and social capital (e.g. Roseland 1999). For me there are parallels between caring for and maintaining the soil within ecosystems (Hill 1989), communities within societies, and the 'shadow' within the self. The tragedy is that not only are we rapidly eroding soils and communities, we are also losing the knowledge and skills and institutional structures and processes that are needed for their ongoing creation and maintenance; and in the process we are also losing our sources of imagination, creativity, intuition and wisdom.

One approach used by many social ecologists to help address such problems involves supporting the formation of 'learning communities' (Senge 1992) and 'collaborative inquiry groups' (Heron 1996). These may then provide the ground within which the needed benign structures and processes can co-evolve. Hunter et al. (1997) have integrated these and other approaches into what they call 'co-operacy', which they regard as the next stage in our social evolution, after autocracy and democracy. Central principles within a co-operacy include caring and sharing, transparency and access, inclusiveness and participation, comprehensiveness, responsibility and proactivity.

'Spiritual' values

The final inclusion in my short definition of social ecology refers to 'spiritual' values. Here I am concerned that spirituality functions as a spontaneous and integrated expression of our core nature, and not as compensation, escape or addiction. For me, spirituality is concerned with the 'rest', the mystery, the unmeasurable wonder and amazingness of it all – from our still largely unknown origins to our unknowable futures. As such, spirituality is related mostly to the 'unknowing' side of the spiral referred to above. It is not something that needs to be explained and organised in great detail. Rather it is one expression of being human – a source for our creativity and openness to learn and relate. In our distressed state, however, many of us have subjected spirituality to the same organising and controlling forces that have been applied to our other social constructions – hence the existence of so many religions, which restrict both life and our ongoing psychosocial evolution. Although claiming to cater to our deep spiritual needs, most religions are more obviously designed and managed to meet the superficial compensatory desires of their constructors, overseers and followers. I believe that such over-organisation of spirituality is robbing so many of us of contexts within which to develop our sense of wonder, so necessary in turn for the development of our values, respect, caring and responsibility.

Conclusion

I believe that the social ecology framework described above can provide us with the breadth and depth of understanding that is needed to carry us forwards to the next stage in our psychosocial co-evolution. To do this we will need to work with rather than against the narrower disciplines, and be more proactive in collaborating with the other metafields, such as those concerned with sustainable development, peace and futures studies.

Notes

1 Based on my 1999 paper, 'Social ecology as future stories: an Australian perspective', published in *A Social Ecology Journal,* 1: 197–208.
2 This view contrasts with that of critical theorists such as Adorno (1955: 67), who consider that domination of nature by humans proceeded domination of humans by humans. I regard both of these behaviours as mutually reinforcing and the phenomenon as an example of negative psychosocial coevolution.
3 The 'Readers' that we used to make available to our students in social ecology aimed to provide key reference materials in many of the areas referred to in this chapter.

References

Adorno, T. (1955; translated 1981) *Prisms*, Cambridge, MA: MIT.

Bateson, G. (1972) *Steps to an Ecology of Mind*, Northvale, NJ: Jason Aronson.

Beck, D.E. and Cowan, C.C. (1996) *Spiral Dynamics: Mastering Values, Leadership, and Change: Exploring the New Science of Memetics*, Cambridge, MA: Blackwell.

Biehl, J. (1998) *The Politics of Social Ecology: Libertarian Municipalism*, New York: Black Rose Books.

Bookchin, M. (Lewis Herber) (1962) *Our Synthetic Environment*, New York: Albert A. Knopf.

Bookchin, M. (1971) *Post-Scarcity Anarchism*, San Francisco, CA: Ramparts.

Bookchin, M. (1980) 'Open letter to the ecology movement', *Rain,* 6 (7; April): 7–10.

Bookchin, M. (1982) *The Ecology of Freedom: The Emergence and Dissolution of Hierarchy*, Palo Alto, CA: Cheshire Books.

Bookchin, M. (1995) *Re-enchanting Humanity: A Defence of the Human Spirit Against Anti-humanism, Misanthropy, Mysticism and Primitivism*, New York: Cassell.

Checkland, P. and Scholes, J. (1990) *Soft Systems Methodology in Action*, Chichester, UK: Wiley.

Clark, J. (1997) 'A social ecology', *Capitalism, Nature, Socialism*, 8 (3): 3–33.

Dale, A. and Hill, S.B. (1996) 'Biodiversity conservation: A decision-making context', in A. Dale and J. Robinson (eds) *Achieving Sustainable Development*, Vancouver, BC: University of British Columbia, 97–118.

deMause, L. (1982) *Foundations of Psychohistory*, New York: Creative Roots.

deMause, L. (2002) *The Emotional Life of Nations*, New York: Other Press (see also, www.psychohistory.com).

Eckersley, R. (1992) *Environmentalism and Political Theory*, Albany, NY: SUNY.

Gruen, A. (1988) *The Betrayal of the Self: The Fear of Autonomy in Men and Women*, New York: Grove.

Gutkind, E.A. (1953) *Community and Environment*, London: Watts.

Harding, R. and Fisher, E. (eds) (1999) *Perspectives on the Precautionary Principle*, Leichhardt, NSW: Federation.

Harman, W.W. and Sahtouris, E. (1998) *Biology Revisioned*, Berkeley, CA: North Atlantic.

Harries-Jones, P. (1995) *A Recursive Vision: Ecological Understanding and Gregory Bateson*, Toronto, ON: University of Toronto.

Heron, J. (1996) *Collaborative Inquiry*, London: Sage.

Hill, S.B. (1980a) 'Letter', *Rain,* 6(8): 3.

Hill, S.B. (1980b) 'Observing stressed and unstressed ecosystems and human systems: Means for recovery and value identification', in *Absolute Values and the Search for the Peace of Mankind*, New York: ICF, 1121–37.

Hill, S.B. (1989) 'The world under our feet', *Seasons,* 29(4):15–19.

Hill, S.B. (1990) 'Cultural methods of pest, primarily insect, control', *Proceedings of the Annual Meeting of the Canadian Pest Management Society*, 36: 85–94.

Hill, S.B. (1998) 'Redesigning agroecosystems for environmental sustainability: A deep systems approach', *Systems Research*, 15: 391–402.

Hill, S.B. (1999) 'Landcare: A multi-stakeholder approach to agricultural sustainability in Australia', in A.K. Dragun and C. Tisdell (eds) *Sustainable Agriculture and Environment: Globalisation and the Impact of Trade Liberalisation*, Cheltenham, UK: Edward Elgar, 125–34.

Hill, S.B. (2003) 'Autonomy, mutualistic relationships, sense of place, and conscious caring: A hopeful view of the present and future', in J.I. Cameron (ed.) *Changing Places: Re-imagining Australia*, Sydney, NSW: Longueville, 180–96.

Hill, S.B. and MacRae, R.J. (1992) 'Organic farming in Canada', *Agriculture, Ecosystems and Environment*, 39: 71–84.

Hill, S.B., Vincent, C. and Chouinard, G. (1999) 'Evolving ecosystem approaches to fruit insect pest management', *Agriculture, Ecosystems and Environment,* 73: 107–10.

Hunter, D., Bailey, A. and Taylor, B. (1997) *Co-operacy: A New Way of Being at Work*, Birkenhead, NZ: Tandem.

Huxley, J. (1964) *Essays of a Humanist* ('Education and Humanism', p132), Harmondsworth, UK: Penguin.

Jackins, H. (1992) *A Better World*, Seattle, WA: Rational Island.

Janov, A. (1971) *The Anatomy of Mental Illness*, New York: Berkley Windhover Books.

Josselson, R. (1996) *The Space Between Us: Exploring the Dimensions of Human Relationships*, Thousand Oaks, CA: Sage.

Lewin, K. (1935) *A Dynamic Theory of Personality: Selected Papers by Kurt Lewin*, New York: McGraw Hill.

Light, A. (ed.) (1998) *Social Ecology after Bookchin*, New York: Guildford.

Mahrer, A.R. (1978) *Experiencing: A Humanistic Theory of Psychology and Psychiatry*, New York: Brunner/Mazel.

Marshall, P. (1992) *Nature's Web: Rethinking Our Place on Earth*, London: Wellington House.

Merchant, C. (ed.) (1994) *Key Concepts in Critical Theory: Ecology*, Atlantic Highlands, NJ: Humanities.

Naess, A. (1989) *Ecology, Community and Lifestyle: Outline of an Ecosophy*, Cambridge, UK: Cambridge University.

Norgaard, R. (1994) *Development Betrayed: The End of Progress and a Coevolutionary Revisioning of the Future*, New York: Routledge.

Raffensperger, C. (1998) 'Guess who's coming for dinner? The scientist and the public making good environmental decisions', *Human Ecology Review*, 5 (1): 37–41.

Raffensperger, C. and Tickner, J. (eds) (1999) *Protecting Public Health and the Environment: Implementing the Precautionary Principle*, Washington DC: Island.

Reason, P. and Bradbury, H. (eds) (2001) *Handbook of Action Research*, London: Sage.

Roseland, M. (1999) 'Natural capital and social capital: Implications for sustainable community development', in J.T. Pierce, J.T. and A. Dale (eds) *Communities, Development, and Sustainability Across Canada*, Vancouver, BC: UBC Press, 190–207.

Rowan, J. (1993) *The Transpersonal: Psychotherapy and Counselling*, London: Routledge.

Rushkoff, D. (1996) *Children of Chaos: Surviving the End of the World as We Know It*, London: Harper-Collins.

Sattmann-Frese, W. and Hill, S. B. (2008) *Learning for Sustainability: Psychology of Ecological Transformation*, Morrisville, NC: Lulu (www.lulu.com).

Savory, A. and Butterfield, J. (1999) *Holistic Management: A New Framework for Decision Making*, Washington DC: Island.

Sazanna, J. (1999) *Understanding and Supporting Young People*, Seattle, WA: Rational Island.

Senge, P.M. (1992) *The Fifth Discipline: The Art and Practice of the Learning Organisation*, Sydney, NSW: Random House.

Shem, S. and Surrey, J. (1998) *We Have to Talk: Healing Dialogues Between Women and Men*, New York: Basic Books (see also, www.wcwonline.org).

Smith, M.J. (1998) *Ecologism: Towards Ecological Citizenship*, Buckingham, UK: Open University.

Solter, A. (1989) *Helping Young Children Flourish*, Goleta, CA: Shining Star.

Stallibrass, A. (1989) *Being Me and Also Us: Lessons from the Peckham Experiment*, Edinburgh, UK: Scottish Academic.

Stettbacher, J.K. (1991) *Making Sense of Suffering*, New York: Dutton.

Watson, D. (1996) *Beyond Bookchin: Preface for a Future Social Ecology*, Detroit, MI: Black & Red.

Williamson, G.S. and Pearse, I.H. (1980) *Science, Synthesis and Sanity*, Edinburgh, UK: Scottish Academic.

Yeomans, P.A. (1958) *The Challenge of Landscape*, Sydney, NSW: Keyline.

2

ATTEMPTING AN INTEGRAL EARTH STORY IN THE NEW CENTURY

Prospects for a twenty-first century education

Edmund O'Sullivan

> Without a meaningful, believable story that explains the world we actually live in, people have no idea how to think about the big picture. And without a big picture, we are very small people.
>
> (Primack and Abrams, 2006)

Introduction

Modern Western culture is probably the first culture to attempt to function without an overriding view of the cosmos. In fact, there has been disenchantment with a larger cosmological context that has developed in modern Western thinking. Frequently, from our Western point of view, we have labelled cultures as retrograde for having a larger cosmology embedded in their mythic structures. At the end of the last century most of the social sciences labelled peoples with mythic interpretations of the universe as primitive. This served more than one purpose. From a scientific point of view, it established Western scientific thinking as superior to the thinking of other existing cultures. The label of primitive also gave European cultures an excuse and apology for their imperialism and colonialism (Said 1993).

We are now living in a watershed period of history comparable, if not even more dramatic, to the major shift that took place from the medieval into the modern world. We are in a transition from a modernist worldview to a postmodern worldview whose characteristics we only vaguely comprehend at this point in time. The educational framework appropriate for movement into this postmodern period must be visionary and transformative, and clearly must go beyond the conventional educational outlooks that we have cultivated for the last several centuries. At this crucial juncture in time, there are compelling reasons to attempt what David Griffin (1988) calls a reconstructive postmodern vision. It must be done with humility and openness. In light of the destructive consequences of the grand narrative of the global marketplace, it is necessary to proceed, with all due caution, with a vision that is comprehensive in scope and magnitude, to counter the destructive totalizing mono-logical system of the market vision. We will need new cultural stories of sufficient power and complexity to overcome the enormous environmental problems that we are facing at this crucial hour. These stories must have the power to activate the possibilities for transforming our world and to reveal to people the role that they can play in this transformative project. The scope and magnitude for such a project cannot be underestimated. The philosopher Arran Gare (1995) speaks with great clarity on these matters pointing out that:

> In order for such stories to 'work', to inspire people to take them seriously, to define their lives in terms of them, and to live accordingly, such stories must be able to confront and interpret the stories by which people are at present defining themselves and choosing how to live in an environmentally destructive way. It is also important to reveal how power operates, and show why those individuals who are concerned about the global environmental crisis are unable effectively to relate their own lives to such problems. The new grand narrative must enable people to understand the relationship between the stories to which they define themselves as individuals and the stories by which groups constitute themselves and define their goals, ranging from families, local communities, organizations and discursive formations, to nations, international organizations and humanity as a whole.
>
> (p.115)

It is one thing to severely criticize our Western conceptions of development and another to try to conceive of education in the absence of an overarching paradigm of development. Therefore, if my treatment of education is to include a conception of development, it will be necessary to articulate a conception of development in a way that it will transcend the limitations of our Western ideas on development and its attendant conception of under-development. What I intend to do in this piece is to set forth a conception of development that I will be calling 'integral development'; and link that concept to the creative evolutionary processes of the universe, the planet, the Earth community, the human community and the personal world. Integral development must be understood as a dynamic wholeness where wholeness encompasses the entire universe and vital consciousness resides both within us and at the same time all around us in the world. Development therefore involves our entry into this larger soul of the world or, as Carl Jung named it, *the anima mundi.*

The fundamental idea of development presumes that all living processes are always in dynamic states of growth, decay and transformation. The idea of evolution and development are different sides of an underlying dynamic. Evolution basically means that all living forms evolve and exceed themselves. Development connotes the dynamic energy that drives the movement of evolving forms. The word growth is used in tandem with development when evolving forms move to higher levels of complexity and integration and can be seen as such. I use the term integral development rather than holistic or integrated development because of the creative dynamic and evolving nature of the processes. The term holistic places an undue emphasis on harmony and integration. My sense of the term integral is that it connotes a dynamic evolving tension of elements held together in a dialectical movement of both harmony and disharmony. An integral model of development will be generative and open-ended, offering an understanding of evolutionary processes that includes a critical role for stress in the transformation of evolving systems. Here, I am drawing on the theory of 'dissapitive structures' presented by Illya Prigogine in initiating a new science of wholeness based on chaos theory (Prigogine and Stengers 1984).

To understand the fundamental nature of Prigogine's theory of dissipative structures one must proceed with the assumption that all evolving forms are open systems whose forms or structures are maintained by a continuous dissipation (consumption) of energy. Dissipative structures are based on what Prigogine calls the principle of order through fluctuation. From an evolutionary point of view, these fluctuations explain irreversible processes in nature and movement through ever-higher orders of life. All systems, including human beings, contain subsystems that are continually fluctuating. All dissipative structures can well be described as an integral system of *flowing wholeness.* Integral development refers to a pattern of linking processes involved in the organization of a system or structure. We can observe that any living system is connected at various points; the more complex the living dissipative structure the more energy is required to maintain its connections. Any system at any given moment can operate in a state of equilibrium or disequilibrium. The continuous flow of energy through a system creates fluctuations within it, many of which are absorbed or adjusted to without altering the system's structural integrity. But if the fluctuations in the system reach a critical level the system becomes sufficiently turbulent so that the old connecting points no longer work. The consequence is that the system commences transforming itself into a higher order with new and different connecting points. The dissipation of energy creates the potential for a sudden reordering of the system. The parts reorder into a new whole and the system escapes into a higher order. Each new level is more integrated and connected than the preceding one and requires a larger flow of energy to maintain it. Every time there is a transformation the system becomes still less stable and, therefore, more susceptible to further change. One can say here that development promotes further development.

Following from these current ideas on evolutionary processes one can now appreciate the need to keep some conception of development even though, in some ways, our Western use of the idea has brought it into disrepute.

Integral development and creativity

The universe has violent as well as harmonious aspects; but it is consistently creative in the larger arc of its development (Jantsch 1984). Evolution does not carry its venture, so to speak, in a placid unfolding of structures. It appears that the processes of evolution are carried out as a dynamic emergence that entails both stability and disequilibrium. From our previous discussion of dissipative systems theory, one understands that when a system reaches beyond its present structure toward new orders of self-organization, it becomes creative in its self-organizing. This creative process of self-organization is called autopoiesis. Autopoiesis refers to the characteristic of living systems to continuously renew themselves, and to regulate this process in such a way that the integrity of the structure is maintained (Berry 1989). It is helpful, at this point, to make the wider connection of our meaning of integral development with the expansive processes of creativity embedded in the unfolding of the universe's story. The entire creative sequence of evolutionary development, from the very beginning, is revealed in the following three basic principles, as set out in one of the earliest works by the cultural historian, Thomas Berry, and the mathematical physicist Brian Swimme (Swimme and Berry 1992): differentiation, subjectivity and communion. These three principles exist as dynamic emergent evolutionary processes and are, at a most fundamental level of definition, the very essence of creativity itself.

When we are talking about contemporary creativity we note that it consists of activating, expressing and fulfilling the universe process, the Earth process, the life process and the human process within the possibilities of the historical moment. Our own historical moment demands that we come to grasp the self-regulating autopoietic processes of our planetary system. As humans, we are now in need of a consciousness that allows us to recognise our own self-regulation within the larger autopoietic processes of the Earth that is our matrix or, if you will, our mother.

Our treatment of the idea of integral development is to imagine it initially within the light of principles of differentiation, subjectivity and communion. The universe did not become a homogeneous smudge, but a world of identifiable and structured beings radiant with inner intelligibility and individual identity. What is essentially implied by the principle of differentiation is the sensitivity to variety in all levels of the evolutionary process. The principle of subjectivity points to the interiority of differentiated processes. While differentiation distinguishes from others, subjectivity gives the interior identity and formation, the inner spontaneity, the indwelling self of every being, its immediacy with ultimate mystery. Subjectivity expresses the autopoietic nature of living things. Out of this inner world of self-regulation comes the freedom that is so minimal in its earlier physical expression that it seems non-existent, although its presence is revealed as the ascending sequence as developments take place. These lead to the variety of living forms and eventually to the human where inner freedom of a high order is attained (Swimme and Berry 1992).

It is within the context of the human that the principle of subjectivity takes on a psychological context. Certainly the domain of phenomenological psychology is most closely connected with aspects of subjectivity and consciousness (Swimme and Berry 1992). Nevertheless, it is appropriate to include the area of cognitive-developmental psychology within our framework of subjectivity. Here we are talking about the seminal work of Chomsky in language, Piaget in cognitive development and Kohlberg in the area of moral development (Sullivan 1990). The problem that we find with all of the above is that there is not a deeply relational quality to their theories; and they collapse the element of community, which is our third fundamental. Thus psychology, as it is developed, currently absolutises subjectivity to the detriment of differentiation and community.

Finally, the principle of communion is pervasive in the emergent evolution of the universe. Already there is by force of gravity, and of other energy bonds, an intercommunion of articulated realities of the universe from the beginning. The atomic structure within itself is a communion of particles. So with the stars in their galaxies, so with the components of the planet Earth, so with the variety of living forms that are interwoven in an enduring web of life, so too with the human. As with differentiation and subjectivity, so with communion which attains its highest expression in human consciousness, in the centre of emotional attraction, and in human aesthetic feeling. We cannot underestimate how important our sense of communion is for the deeper needs of our very existence. It is fundamentally important to understand what the loss of communion means for us in our daily living. The intricacy of our personal world is embedded in community. Much of our existence finds ultimate fulfilment in relatedness. We can see this in the intricate mating rituals the natural world has invented. So much of the plumage, coloration and dance and song of the world comes from our relationships of true intimacy. The energy that we and other animals bestow on this work of relatedness of being, the attention that we give to our physical appearance alone reveals the ultimate meaning of the communion experience (Swimme and Berry 1992).

The loss of relationship and the consequent alienation is a state that all of us experience, but when it is sustained it is a terrifying experience. To be locked up in a private world, to be cut off from intimacy with other beings, to be incapable of entering the joy of mutual presence is a state that in previous times was akin to a sense of damnation. Today we refer to this state of excommunication as alienation. This alienation is cultivated when there is an extreme emphasis on individuation in the differentiation process, or when the basis of our subjectivity is undermined (Swimme and Berry 1992).

The former is characterized as egocentrism, while the latter is characterized as autism. What we perceive today in contemporary Western values is the embodiment of alienation. Thus in a world of vast diversity, with its promise of an equally vast diverse relatedness, modern humans find themselves shut up in a world of egocentrism and autism, incapable of any deep or sustained contact with the world outside the self. Even when there is a sense of community, the community is limited to the human; the world outside the human is excluded. This truncated sense of community is labelled anthropocentrism. Its subjective correlate is *individualism,* the alienated exaggeration of the differentiation process when considered independent from *communion.* I consider it very important to address the historical development of the idea of individualism because of its central value in modern consciousness. In addition, we must assess, at the personal level, the profound effect of the value of *individualism* within the context of disenchantment. The modern definition of the individual as an autonomous social unit is a product of the consensus achieved by Western liberal social theory. In liberal social theory, individuals are seen as separate autonomous monads that are unique unto themselves. The primary position of the 'state of nature' is characterized by Hobbes as solitary, where the individual contracts socially out of fear of survival. The creation of society is based on a contractual arrangement of separate individual entities, who in Hobbes words were social atoms. This atomization is characteristic of the liberal idea of individualism, and the downside is now being felt at all levels of cultural life. The self-encapsulated individual that we have just described has profound implications for the loss of cosmological sense that links individuals to the wider community and subsequently to the universe itself.

Human emergence and integral development

We must understand our human presence on this Earth as an integral part of the evolutionary processes of the universe, and of the evolution of the planet Earth. Thomas Berry's seventh principle articulates human emergence as follows:

> The human emerges within the life systems of the earth as that being in whom the universe reflects on and celebrates itself in a special mode of conscious self-awareness. The human is genetically coded toward a further trans-genetic cultural coding invented by the human community in a remarkable diversity in the various regions of the earth. This diversity of cultural elaboration in the various peoples of the earth is communicated to succeeding generations by both formal and informal educational processes.
>
> (Swimme and Berry, 1992)

Humans emerge in the processes of evolution as a unique species of planet Earth. In broad evolutionary terms, humans are a 'one-time, one-place' species. In saying this we acknowledge that the human emerges at a specific time in the development of the universe and also is restricted to a particular place. When we say place we acknowledge that humans are uniquely creatures of the planet Earth. This means that in the unfolding of the universe that humans emerge at a specific time and place. We, as a species, are peculiar creatures of the planet Earth. Our existence in time is historical time. We are the creation, in time, of the Earth's unique unfolding. While we say that the Earth is a 'one-time, one-place' event in the evolution of the universe, we can also say that the human species is unique in its time and place in the cosmic unfolding. Not until the Earth is saturated with life in such a dazzling complexity of form could something like the human arise. The human being is not only an Earthling through and through in the sense that every molecule and every organ and every physiological action pattern has been woven out of the living strands of our planet; it is also true to say that the human is a time-specific expression of the Earth in its luscious existence four million years ago, when the continents had reached an eco-systemic complexity never before attained, and for all we know never again to be experienced. Out of a paradise of beauty and elegance, the Earth seemed to have surpassed itself again in an ecstatic self-display that we call the human (Spretnak 1991).

The Earth is the central locus of the evolutionary process within our planetary system. It appears that the possibility for the Earth was, first, contained in the evolution of the galaxies and the elements; second, we see its actual birth in the evolution of the solar system; third, it is the container for life in all its variety; and fourth, it generates the evolution of consciousness and the cultural developments of the human order.

It is especially important in this discussion to recognise the unity of this total process from that first unimaginable moment of cosmic creation until the present. This unbreakable bond of relatedness is increasingly apparent to scientists, although it ultimately escapes scientific formulation or understanding. By virtue of this relatedness, everything is intimately present to everything else in the universe. Nothing is completely itself without everything else. This unity prevails over the boundaries of space and time. The universe is both communion and community. We ourselves are that communion in a special mode of reflective awareness.

The invention of human community

Instinctively, humans perceived themselves as a mode of being of the universe, as well as distinctive beings in the universe. This is what we mean when we say existence. The emergence of the human was a transformation moment for the Earth as well as for the human. As with every species, there was a need to establish its niche, a sustainable position in the larger community of life, the need for food and shelter and for clothing. There was need for security, the need for family and community context. This need for community was quite special in the case of humans since humans articulate a special mode of being, a capacity for thought and speech, aesthetic appreciation, emotional sensitivities and moral judgement, all uniquely human. These combine in a cultural shaping that establishes the human in its specific identifying qualities.

Whatever the cultural elaboration of the human, its basic physical as well as psychic nourishment and support came from the surrounding natural environment. Human society in its beginnings was integral with the larger life society, and with the larger Earth community composed of all the geological as well its biological and human elements. Just how long this primordial harmony of the very early period endured we do not know beyond the last hundred thousand years of the Paleolithic period. Some ten thousand years ago, the Neolithic and then the classical civilisations came into being. It must suffice here to indicate that with the classical and generally literate civilisations of the past five thousand years, the great cultural worlds of the human developed, along with vast and powerful social establishments whereby humans became oppressive and even destructive of other life forms.

We must be very aware at this time of our Earth history that the human presence on this Earth has taken on a variety and diversity of forms through its brief evolutionary history. We seem to be an open-ended species that has the capacity for invention of ourselves along a number of different paths. The reason that I am stressing the inventiveness of human–Earth relationships is to have the reader sensitised to the fact that we have invented and reinvented ourselves many times during the course of our brief history on this Earth. Today we need to be able to project a new vision of ourselves in relation to our presence on this Earth. In our modern era, we have invented tools and devices that are bringing us to a disastrous scenario within the Earth context. Our hope, in this moment, is that we will be able to call upon our capacities of invention and creativity in order to forge a presence that is mutually enhancing in terms of integral human–Earth relationship.

That the human emerges out of the planetary process and is dependent on the planetary process in its basic laws of development is quite evident, especially during the earlier phases of the human enterprise. More so than other life forms, the human had to discover its proper identity and the role within the Earth process and within the life community, both by instinct and by thought processes. Humans needed to frame for themselves some sense of the natural world, how it came into being, how it functioned, and what the proper role of the human might be. These perceptions were presented in stories that narrated the creation of the universe and the events leading up to the present. Within this context the human could proceed with its self-identification, and establish patterns of action coordinated with the surrounding universe.

The difference in these stories resulted from, and further enhanced, the distinctive life styles whereby the various peoples related to each other, to the natural world about them, and to the numinous powers that provided their ultimate explanation for the emergence of the universe and for its patterns of activity. Through their various myths and through their rituals, based largely on the seasonal transformations of the natural world, the various societies established their functional presence within the surrounding universe. The differences in these myths and rituals reflected the diversity of life experience of the various peoples, and the special features of the geographical region they inhabited beyond the human were that they were objects to be manipulated and exploited for human designs. Thomas Berry speaks of the human as in a state of autism. He maintains that we have lost our sense of the awesome mystery surrounding us, and we no longer understand the voices speaking to us from the surrounding world.

The human need for the diversity of the natural world

The disenchantment from the natural world that I have identified with the modern project has resulted in a dynamic process that has put human beings in an adversarial relationship to the natural world. We have, in the modern period, centred all our sense of value on the human historical project. What we are currently experiencing is a dynamic of shrinking from the world beyond the human, either through a belligerent distancing or a wall of indifference. This outcome has been one of the central components of the modern project. We now are beginning to understand the costs that this project of modernism entailed. The natural world beyond the human has become seriously degraded, and we are now living in a clearly dysfunctional relationship with the natural world. Our sense of the natural world of the planet has left us in a state of insensitivity to the natural world in the deeper emotional, aesthetic, mythic and mystical communication. Just as autistic persons are enclosed in themselves so tightly that they cannot get out of themselves and nothing else can get in, so we are presently enclosed in our human world that, as a society, we have lost our intimacy with the natural world. Our modern education in industrial societies has educated us away from intimacy. It has initiated a journey into estrangement. We have been taught to see ourselves as separate and detached from the natural world. When we talk of our existence, we speak of it as standing out from and separated from the universe and the natural world. Looked at from this perspective, human consciousness is conscious insofar as it is seen as separated from the universe and the natural world. When one considers the worldview of Western thinking, we perceive the world outside our consciousness as silent and inert: dead matter to be manipulated and controlled at our fancy. The world outside human consciousness is an object to be used at our species' discretion, and within the terms of our own needs and preferences. The only intimacy within this worldview is human intimacy. Here we are locked into anthropocentrism. The only other voice besides our own is the human; all other aspects of the natural world outside the human are silent and speechless. Because we imagine the universe outside the human as having no voice, it follows that we have no capacity for intimacy with the natural world. We are not aware that we are living on the surface of things. This is part of the arrogance of Western education. We look at cultures that profess an intimacy with the world outside human consciousness as primitive and underdeveloped. We see this clearly in the case of indigenous peoples. Their intimacy with the natural world, their closeness and reverence for the animal and plant world are seen as retrograde within the modern temper. What we in the modern Western scientific tradition have failed to recognise is our very limited capacity for communion with the world outside of the human. Our Western educational tradition has trained us to think only of the human world as subjects. But what if the universe is a communion of subjects and not a collection of objects? This sense of communion with the natural world must now become fundamental to our educational experience.

We must now move from anthropocentric to a biocentric sense of reality and value. This begins in accepting the fact that the life community, the community of all living species, is superordinate in value; and the primary concern of the human must be in the preservation and enhancement of this larger life community. When we consider that the universe is a communion of subjects and not a collection of objects we commence to hold as sacred the deep interiority of all aspects of being. As Charlene Spretnak (1991) puts it, 'humans are not the only subjects in the universe'. She indicates that it is within our sensibilities to envision the universe itself as a grand subject. A greater engagement with the natural world in terms of its deep subjectivity opens up a new sense of intimacy. When intimacy with the natural world is cultivated, we begin to see a differentiated consciousness to the world outside the human. Sensitivities to the animal and plant world open up a consciousness that brings about a sensitivity to the deeper rhythms of the biotic world. Humans now are able to enter a relationship with the natural world that honours the deep subjectivity and interiority of all aspects of reality. With this wider differentiated consciousness there is the expanded capacity to comprehend all of reality as both a different and a subjective presence. With this expanded sensitivity and awareness we commence to develop an inner poise that allows a deep relational insight into everything that we may experience in and around us. This would be grounded in awe and respect for the larger biotic community: the web of life

References

Berry, T. (1989) 'Twelve principles for understanding the universe and the role of the human in the universe', *Teilhard Perspective*, 22: 1, July, 214.

Gare, A, (1995) *Postmodernism and the Environmental Crisis*, New York: Routledge.

Griffin, D. (1988) *The Reinchantment of Science: Postmodern Proposals*, New York: SUNY Press.

Jantsch, E. (1984) *The Self-organizing Universe: Scientific and Human Implications of the Emerging Paradigm of Evolution*, New York: Pergamon.

Prigogine, I. and Stengers, I. (1984) *Order Out of Chaos: Man's New Dialogue with Nature*, New York: Bantam.

Primack, J. and Abrams, E. (2006) *View from the Center of the Universe*, Riverhead, NY: Penguin.

Said, E, (1993) *Culture and Imperialism*, New York: Alfred A Knopf.

Spretnak, C. (1991) *States of Grace: Recovery of Meaning in the Postmodern Age*, San Francisco, CA: Harper.

Sullivan, E. (1990) *Critical Psychology and Critical Pedagogy*, New York: Bergin and Garvey.

Swimme, B. and Berry, T. (1992) *The Universe Story*, San Francisco, CA: Harper.

3

THE PEDAGOGIC 'STING'

Social ecology and narrative imaginal pedagogy

Peter Willis

Introduction

This chapter begins with a reflection on a biblical story that I remember from student days, not so much for its specifically religious provenance as for its dramatic and pedagogic 'sting'. The story (2 Sam, 11) is that of King David, one of the early Jewish heroes in the Old Testament, who falls in love with Bathsheba, a married woman. He engineers the death of her soldier husband by putting him where the fighting is fiercest in one of the local battles, and then marries her. The prophet Nathan is then sent by the Lord to meet with King David. The king receives the Lord's prophet who tells him this special story. It is about a rich man who had large flocks and a poor man who had one small ewe lamb. When a traveller came, the mean rich man took the poor man's only lamb to cook for the traveller.

King David, usually a warm-hearted and honourable man, becomes upset and angry. He bursts out that this cruel, rich man has shown no pity and deserves to die. At this point Nathan says the revealing words: *Thou art the man*. Guilelessly caught up in the story with no opportunity to think of excuses and rationalisations, the king sees the enormity of his action revealed in Nathan's metaphor of selfish and merciless greed. Nathan's storytelling pedagogy not only proves the king wrong but reaches the king's heart so that he repents, makes reparation and turns away from selfishness to take up his previous life of duty and obedience to the Lord in a chastened and clearer way.

In this chapter I suggest that the message, story or parable used by the prophet Nathan to capture the imagination and move the heart of King David was a form of *narrative imaginal pedagogy*, and that versions of this approach can be useful in promoting the values and practices of social ecology. Before exploring this approach in detail, I firstly introduce the idea of social ecology and its educational dimension, then suggest a heuristic fourfold idea of human learning drawing on ideas of John Heron. I then focus on the second or imaginal mode of knowing and its links to human mythopoesis or deep storytelling. Finally, I suggest that the pedagogy of social ecology needs to have a significant and delicate imaginal dimension for which narrative imaginal pedagogy can be an appropriate nurturing approach to foster a powerful and attractive vision of social ecology in human communities.

Social ecology and its learning agenda

Social ecology refers here to ways of purposively seeking to shape and conserve the interdependent processes characterising social groups and their members within their physical and social environment. The general aim is to encourage positive outcomes in the middle of processes of growth and decline, wounding and healing, conflict and reconciliation in and through time. This enterprise has idealistic dimensions and is pursued knowing the difficulty of attempting intervention into such systems while seeking to safeguard life, autonomy and resilience. Social ecology draws originally from the ideas of the North American anarchist, Murray Bookchin (1993), and his holistic social vision. In Australia a broad humanistic version of social ecology has been taught at the University of Western Sydney. Stuart Hill, one of its key protagonists, writes that social ecology:

> emphasizes actions and reflective practice that integrate personal, social, political and environmental concerns and possibilities. End goals include wellbeing and health, in the broadest sense, equity and social justice, and the fostering of mutualistic and caring relationships, personal meaning, organizational learning, co-evolutionary change and ecological sustainability.
>
> (2000: 1)

In this context, although the dimensions of social ecology are not fixed as an applied ideology, they not unlike some forms of spirituality (cf. Willis and Morrison 2009: 5) pursued in a range of receptive and hostile environments. They can be seen to have three elements: belief systems, fostering practices and applications in everyday life in the development of social ecological practices in different circumstances. These three elements need to be highlighted and addressed in different ways.

What I have been trying to do in this paper is to map out dimensions of an appropriate respectful pedagogy that might promote the holistic and inclusive social ecological vision. I am interested in exploring the kind of human learning required to generate and sustain a social ecology among people in a nation or region. The social change I envisage, and which I believe is integral to social ecology, is one of flexible accommodation and inclusivity rather than conflict and the defeat of enemies. The curricular challenge is thus to explore what kinds of learning environments could promote social ecology and how these could be created. The learning complexity and challenges of social ecology as a holistic agenda seem to need multiple pedagogies fine-tuned to different kinds of knowing.

The learning groups under consideration here were not restricted to the formal educational classrooms at a university or technical college or workplace training site. Social ecology wants to connect and engage with any group of people coming together for common interests, such as sport, expressive culture, artistic and leisure pursuits, religious interests and of course learning and social support. Every such gathering develops working values through the exchanges that underpin their activities.

The social ecological vision invites members of all kinds of groups to reflect on and seek to nurture their group as a social and cultural system with capacity to enrich or, of course, diminish their lives and the social and physical environment they inhabit. In many cases, well run clubs and community groups with tactful and energetic leadership manifest many social ecological elements under different names. The contribution of a social ecological vision is its capacity to raise awareness about the multiplex energies of human social life that are activated in one way or another in the day-to-day lives of organisations. This includes social relations and leadership, as well environmental action in the use of water and food, power and the like. Social ecological awareness looks to raise awareness of and support for all the systems in a community so that, where necessary, some can be healed, repaired and modified.

A key underpinning learning process seemed to be needed to trigger and sustain a desire for holistic engagement by focusing on ways of evoking and developing three dispositions that, according to Hill (2000: 2), are at the core of social ecology:

- Contextual awareness: a need to become powerfully and critically aware of the social ecological contexts in which one finds oneself.
- Self-reflection: sensitive to one's own life stance in comparison to the life stance required for social ecology.
- Predisposition to care: where the world and its human inhabitants in large and small arenas become part of one's concern.

The question is then to develop an appropriate pedagogy that can address these dispositions while maintaining respect for the autonomy and self-determination of learners.

Such a holistic agenda needs to speak to the various elements of human awareness and choice. This requires some preliminary ideas of the dimensions of human knowing and learning. I have found the heuristic ideas of John Heron, developed nearly two decades ago, relevant and useful to the quest.

Four modalities of knowing and social ecology

John Heron (1992: 14) suggests that a heuristic way to order knowing and learning experiences is to envisage them as having four modes, each leading to the next in a circle. The circle begins with raw, embodied sensation/connection to things and experiences. This experience/connection gives way to direct image experiences, which are named as the imaginal or metaphoric mode of knowing. The next mode is reasoning that is analytical and critical. This leads finally to the praxis modes of knowing that come from reflective purposive action on the world.

For Heron (1996: 33), these four modes of knowing and learning can be distinguished to a greater or lesser extent in the purposive activities of everyday human life. *Initial sensation/connection action* is evident when we become directly aware of all kinds of things and inner states; *imaginal action* is evident in our intuitive grasp of aesthetic patterns in art forms and stories; *reasoning action* is expressed in critical and logical appraisal, while *practical action* is evident in the direct knowledge of practice. According to Heron, each mode of knowing evokes the next. Thus the sensation mode evokes the imaginal, the imaginal the critical, and the critical the praxis.

This chapter explores ways in which the ideology of social ecology as an active and applied practical ideology with the three elements mentioned above, beliefs, fostering practices and applications in everyday life, could be promoted in Heron's four modes of knowing and learning. Thus initial sensation/connection modalities can be linked to what can be called *awareness* social ecology, imaginal modalities with *mythopoetic* social ecology, rational modalities with *critical* social ecology and reflective action-based modalities with *praxis* social ecology. The following expands these a little, but the focus is on the second, the mythopoetic, with its specific links to narrative, imaginal pedagogy.

Four dimensions of social ecology

The *awareness* dimension of social ecology seeks to celebrate and shape human awakeness of one's personal orientation to and connection with self, with other humans and with the wider non-human world. Since this mode of knowing is linked to *initial sensation/connection* prior to imaginal, critical and praxis forms of knowing, it is linked not so much to ideology development, the first of the three elements of social ecology practice mentioned above, but more to fostering practices to develop 'awakeness' to the world. Awareness and awakeness dimensions of knowing tend to have a strong embodied dimension and to be a significant foundation for the application of social ecology in other forms of human knowing and learning.

Mythopoetic social ecology can be said to draw on James Hillman's (1976, 2000) ideas of human imaginal knowing and learning, which he developed through his interpretation of Henri Corbin's (1969) translations and expositions of Ibn 'Arabi, the Sufi mystic. According to Hillman, the mythopoetic or imaginal mode of knowing and learning 'provokes, delights, confuses, tantalises' (1976: 158). It can 'start us imagining, questioning, going deeper'. This kind of knowing and learning cannot be generated by direct instruction since it is not about being 'right', but about being 'interesting' or 'enchanting'. Its agenda is one of invitation and attraction rather than the conviction of propositional truth. Mythopoetic knowing in social ecology serves to enhance the aesthetic quality and existential enrichment of life. The imaginal knowing of mythopoesis is not the same as the workings of the imagination, which are much broader and full of possibilities that may or may not be grounded in reality (cf Hillman 1981; Boyd and Myers 1988;

Dirkx 1998, 2000). Imaginal processes are linked to the way people consciously or unconsciously value things and experiences, seemingly instinctively and often without full awareness. These, according to Hillman, are supported by 'generative' or 'mythic' images, images that hold the imagination and move the heart, the influence of which is somehow present deep in the person's psyche.

Following Hillman's imaginal knowing and Macdonald's (1981) related idea of mythopoesis, Bradbeer (1998: 45) suggests that people's personal myths, the ideals and visions that inspire their life and work, are knowingly or unknowingly juxtaposed against the powerful so-called 'archetypes' of the unconscious mind by this mythopoetic process. Mythopoetic forms of knowing complement logical rationality. They are evoked when people, reflecting on important moments of their life and dreams and focusing on central images appearing in them, seek to develop a 'way of talking with the image and letting it talk' (Hillman 1977: 65). This mythopoetic arena is also an intensely aesthetic one that moves in the human world of beauty and ugliness, of taste and connoisseurship. The work of a social ecology that respects the mythopoetic seeks to create aesthetically arresting mythic narratives around events and experiences of life shaped by the social ecology's belief systems. Mythopoetic processes tend to serve social ecology's ideology and fostering elements mentioned above. When presented in evocative textual processes – poetry, drama and imaginal writing – their impact is heightened further.

The *critical* dimension of social ecology, which has traditionally been one of its significant elements, particularly in challenging exclusively modernist and positivist ideals, draws on the human cognitive capacity to analyse and appraise. Social ecology needs to incorporate rational approaches to foster a critical and ethical capacity to guard against over-credulity on the one hand and unethical practices on the other. The critical dimension of social ecology stresses logical clarity in articulating ideals and their implications, the appropriateness of fostering practices and the usefulness and ethical value of the chosen life applications.

The fourth dimension of social ecology, 'praxis', adds a touch of grounded realism to initial awareness, mythopoesis and critique. Freire (1970, 1973) used the word 'praxis' to refer to the process of putting ideas into action attentively and reflectively. Praxis in social ecology adds a grounded and embodied voice to human dreams and plans.

What I am most concerned to explore in this chapter are narrative versions of imaginal pedagogic approaches that seek to strengthen the mythopoetic dimensions of social ecology. As has been pointed out above, mythopoesis is the creation of meaning from the innumerable myths, stories and images that enrich human living. Thus, mythopoesis is essentially 'the curriculum of the soul'. Learning here is not restricted to the classroom, but to those human exchanges and experiences in which, directly or indirectly, learning is evoked. Mythopoetic learning can occur not only in formal educational contexts, but at moments of stress, challenge or change in life circumstances, whether experienced and reflected upon alone or in the company of a friend, mentor or teacher.

Portrayal and parable: Imaginal pedagogy in social ecology

The question of appropriate pedagogy for social ecology emerged at a 'place writers' workshop held in autumn 2004 at the Far South eco-camp near Dover in south-eastern Tasmania. About forty poets, writers and environmental scientists spent the Easter long weekend reflecting, walking, chatting and performing around the theme of ecological life and a sense of place. I was interested in the various forms of ecological consciousness of participants, some of whom, with more scientific and empirical background and culture, were drawn to the ecology of biological sustainability and biodiversity. Others with a more humanistic life stance were interested in the ecology of human collaboration on all levels of awareness – ecology of production, creativity, community, reconciliation and regeneration. They celebrated their engagement with place and each other with excerpts from essays, and stories and poems of place. I was struck by the inclusive style of several participants who seemed to cross and draw together the multiple themes of social ecology – love and care for the earth as an ecosystem and love and care for the humans in that ecosystem. What made these people significant was the breadth and depth of their awareness and their integrity and conviviality. I felt others noticed it as well.

Some place writer eco-activists were confronted by contradictions in their own life and action narratives. One of the most powerful of these was the question of appropriate ways to campaign for social ecological principles around the environment and social justice. As the conference went on, a couple of activist participants who had angrily used language like 'eco-destroyers' or 'idiots who can't see that the ecosystem is in trouble' to describe the group who did not espouse their views, began, under the unspoken collaborative and inclusive culture of the event, to feel a kind of 'thou art the man' feeling that King David had felt when prompted by Nathan. They mentioned that they had become aware of the rigidity and aggression in some of their comments as they blamed and dismissed opponents. Inviting people to consider change became revealed as radically different from preaching or attacking them and their ideas. Some also began to perceive that for people to become interested in embracing real and possibly uncomfortable change, preliminary fellow feeling was needed, a kind of shared conviviality within which differences could be listened to and considered without too much instant acrimony.

It seemed to me that, while there might be some relevance to the other three modes of knowing, awareness, critical and praxis, these revelatory confronting stories seemed to connect powerfully with people's mythopoetic mode of knowing and their own deep self story. The narrative pedagogy that could generate such deep response, referred to here as *narrative, imaginal pedagogy* is the main theme of the last section of this chapter.

Imaginal pedagogy for social ecology

The idea of imaginal pedagogy, as understood here, is one that seeks to evoke a 'listening' rather then 'resistant' or 'combative' response from the learners that can be facilitated by the credibility and style of the educator and the appropriateness of the *image* and *story* employed. Image and story refer to linked and complementary acts of portrayal and narrative pedagogy. I have written about portrayal pedagogy (cf. Willis 2008), but the pedagogic impact of narrative pedagogy is an addition.

Portrayal imaginal pedagogy provides an evocative picture of the character of the archetypal exemplar with which learners can identify or be repelled and be lead to use a kind of searching learning mirror: could I really become like that and do I want to? *Narrative* imaginal pedagogy uses stories of specific chosen activities to reveal the implications and effects of moral choices with which the learners can identify, carried out by a person who is perceived to be not too 'other' to the learners. In this case it is the morality of the behaviour or activity that is revealed and which learners are invited to reflect on and, of course, extrapolate to their own circumstances when they face similar moral dilemmas.

Linking back to King David at the beginning of this chapter, the pedagogy of the prophet Nathan can be called narrative imaginal pedagogy. His narrative of merciless oppression of the weak is enriched by the credibility and style of the storyteller – the prophet of the Lord – and the power of its visual images – the poor and unprotected man and the one small ewe lamb, a powerful symbol of helplessness and potential fecundity. The intertwined authority of the teller, image and story are perfectly calculated to generate firstly some preliminary respect for the storyteller prophet and then a strong empathetic and protective response to the story in the usually warm-hearted and noble David whose protective instincts were evoked together with his usually passionate sense of fairness and justice. Thus this symbol-filled story uttered by a significant messenger creates a strong, imaginal pedagogy that predisposes the king to hear and listen deeply to a more direct version of the prophetic message that Nathan then delivers. This then generates a deep, mythopoetic response in the King leading to a radical conversion.

The 'conversion' experiences of some of the doctrinaire participants in the place writers workshop seemed also to have been generated by persuasive stories of convivial rather than combative versions of social ecology and, of course, their direct application in the courtesies and exchanges of the camp. These are more recent examples of learners invited into reflective learning by the power and sweetness of an evocative narrative.

The question is then how such storytelling can be developed as a pedagogic form, and that is the question for the final section of this chapter. The following section explores the nature of narrative and story and their capacity to evoke meaning in a number of ways. This leads to a brief comment on the way evocative stories can be used in imaginal pedagogy.

Narrative imaginal pedagogy for social ecology

Frank (2000: 354) suggests that narrative is a more abstract term for the structure of a story rather than its full reality, which is contained in the idea of story. For him stories reaffirm and reconstruct relationships; they can provide a kind of healing; they are more than data for analysis and they are told to be heard, to be listened to, to capture the imagination and move the heart and to find others who will answer their call for a relationship (cf. Frank 1995, 2004). For him and Coles (1989), story has a pedagogic dimension in its agenda to shape and create relationships. Bochner (2002: 80) suggests that stories have a number of common elements. First of all people are represented as *characters* in the story. Secondly there is some kind of *plot* with a critical moment that resolves the dramatic tension. The third element is *time*. Stories place things in temporal order. Finally, stories have to have some kind of *point* that can often, but not of course always, have a pedagogic dimension.

It was this last element that received considerable exploration by Baumeister and Newman. They distinguished (1994: 679) two general categories: firstly stories aimed at *affecting* listeners in some way, and secondly stories whose main objective is as a way of *making sense* of experiences.

Reflecting on the first category, those stories aimed at 'affecting listeners', they identified four general motives for storytelling: obtain rewards, validate identity claims, pass on information and entertain. This is a useful preliminary categorisation that is not understood to be exclusive since one story may have several agendas. It is useful in narrative research to be aware of the kinds of narrative motivations being pursued by storytellers, as well as the kind of knowledge they are seeking to create. Reflecting on the second category, those stories concerned with interpreting and making sense of experience, Reissman (1993: 2) spoke of this kind of narrative research as a way that people 'impose order on the flow of experiences to make sense of events and actions in their lives'.

Stories used in narrative imaginal pedagogy will often have these affective and rational elements, but the aim is to capture the imagination and move the heart of the listeners. Narrative imaginal pedagogy has strong links to Heron's second mode of knowing, the mythopoetic, mentioned earlier. The stories of narrative imaginal pedagogy are largely concerned with awakening and shaping mythopoetic forms of knowing with their links to deep imaginal and emotional responses. They have links to ancient forms of rhetoric like the prophetic narrative, which was apparent in the story of King David at the beginning of this chapter, and the parable.

Narrative mythopoesis: The storyteller, story and telling

Narrative imaginal stories carry a certain gravitas as contributing to mythopoetic life. They are shaped and their authority endorsed by the teller, who needs to have or claim some narrative authority over the listeners. This can be because of the office of a person such as priest, professor or judge, or it can be a claimed authority from a person playing an authoritative role in a drama. When used as pedagogy, narrative imaginal stories need to be told with as much leverage and credibility as possible in order to endorse the dramatic invitation to another but still relevant world.

One of the key elements in the stories of narrative imaginal pedagogy is that they have to be *mythopoetic*, which means they need to resonate with great 'mythic' themes in human life: birth and death, youth, maturation and decline, war and peace, enmity and friendship, love and sexuality, conflict and resolution, work and achievement, work loss and anomie. Under most circumstances, you can't have a narrative imaginal story about a mundane shopping excursion or going to the football – unless of course such events for some are not mundane but are infused with excitement and tribal passion. The themes of murder and adultery in the biblical story are intensely mythopoetic, as are the themes of exclusion and hostility in the social ecological conversion narratives of the environmental activists who moved from righteous and myopic warriors to convivial and inclusive ecological campaigners.

A second element is *appropriate literary artistry* since narrative imaginal stories seek to capture the imagination of the listeners and maintain a kind of enchantment and suspension of judgement. The audience has to feel and be caught up in the invitational undertones of different kinds of imagery and media.

A third element is *dramatic form*. In some cases the dramatic form can be the simple well-crafted text. In others, the imaginal, pedagogic narrative with its tacit challenges is given aesthetic strength by music and poetry and drama. As Hamlet (Hamlet Act 2, Scene 2) said in an aside to the audience before the performance of a play in the royal court, which he had modified for his purpose:

> *The play's the thing wherein I'll catch the conscience of the King.*

Dramatised stories were used by Morey (2010) and by August Boal (1992, 1995) and collaborators in his work on the 'drama of the oppressed'.

A fourth element is *delayed and dramatic denouement*. Narrative imaginal pedagogy seeks to create dramatic tension and delayed resolution of the themes and plots in the story being told. This is where the tacit element in this kind of narrative imaginal pedagogy emerges. The element of surprise is designed to increase the impact of the contradictions that are hinted at originally and perhaps made visible during the body of the play, between what people say they do in following espoused customs and beliefs and what they may actually do from time to time without being consciously aware of deviation from espoused ideals.

Imaginal narrative pedagogy, with its various cultural vehicles, is designed to catch the conscience of others similarly placed and to create a clarifying mythopoetic and moral experience to sharpen and renew a blurred, moral stance.

Conclusion

This chapter has introduced the stories and images used in imaginal narrative pedagogy as appropriate ways to promote a thirst for the ideals and practices of social ecology. It has briefly explored the democratic ideology in social ecology, the fourfold nature of human knowing and learning and the importance for promotional pedagogy in social ecology to promote it's ideals in non-intrusive but powerful ways.

References

Baumeister, R.F. and Newman, L.S. (1994) 'How stories make sense of personal experience: Motives that shape autobiographical narratives', *Personality and Social Psychological Bulletin*, 20: 676–90.

Boal, A. (1992) *Games for Actors and Non-actors*, London: Routledge.

Boal, A. (1995) *The Rainbow of Desire: The Boal Method of Theatre and Therapy*, London: Routledge.

Bochner, A. (2002) *Ethnographically Speaking: Authoethnography, Literature, and Aesthetics*, Walnut Creek, CA: Alta Mira Press.

Bookchin, M. (1993) 'What is social ecology?' in M.E. Zimmerman (ed.) *Environmental Philosophy: From Animal Rights to Radical Ecology*, Englewood Cliffs, NJ: Prentice Hall.

Boyd, R. and Myers, J. (1988) 'Transformative education', *International Journal of Lifelong Education*, 7: 261–84.

Bradbeer, J. (1998) *Imagining Curriculum: Practical Intelligence in Teaching*, New York: Teachers College Press.

Coles, R. (1989) *The Call of Stories: Teaching and the Moral Imagination*, Boston, MA: Houghton Mifflin.

Corbin, H. (1969) *Creative Imagination in the Sufism of Ibn 'Arabi*, Princeton, NJ: Princeton University Press.

Dirkx, J.M. (1998) 'Knowing the self through fantasy: Toward a mytho-poetic view of transformative learning', in J.C. Kimmel (ed.) *Proceedings of the 39th Annual Researcher Education Research Conference*, University of Incarnate Word and Texas A&M University, San Antonio, TX, 137–42.

Dirkx, J.M. (2000) 'Transformative learning and the journey of individuation', *ERIC Digest*, 223, 1–2.

Frank, A. (1995) *The Wounded Storyteller: Body, Illness and Ethics*, Chicago, IL: University of Chicago Press.

Frank, A. (2000) 'The standpoint of story teller', *Qualitative Health Research*, 10(3): 354–65.

Frank, A. (2004) *The Renewal of Generosity: Illness, Medicine and How to Live*, Chicago, IL: University of Chicago Press.

Freire, P. (1970) *Pedagogy of the Oppressed*, London: Penguin.

Freire, P. (1973) *Education: The Practice of Freedom*, London: Writers and Readers.

Heron, J. (1992) *Feeling and Personhood: Psychology in Another Key*, Newbury Park, CA: Sage.

Heron, J. (1996) *Co-operative Inquiry: Research into the Human Condition*, London: Sage.

Hill, S.B. (2000) *Social Ecology* (Residential Notes), Richmond, NSW: University of Western Sydney.

Hillman, J. (1976) *Re-visioning Psychology*, New York: Harper & Row.

Hillman, J. (1977) 'An inquiry into image', *Spring*, 62–8.

Hillman, J. (1981) *The Thought of the Heart*, Dallas, TX: Spring Publications.

Hillman, J. (2000) 'Peaks and vales', in B. Sells (ed.) *Working with Images*, Woodstock, CT: Spring Publications.

Macdonald, B.J. (1981) 'Theory, practice and the hermeneutic circle', in B.J. Macdonald (ed) (1995) *Theory as a Prayerful Act. The Collected Essays of James B. Macdonald*, New York: Peter Lang Publishing.

Morey, O. (2010) *Loving Care for a Person with Dementia: From Phenomenological Findings to Lifeworld Theatre*, unpublished PhD thesis, Bournemouth, UK: University of Bournemouth.

Reissman, C.K. (1993) *Narrative Analysis*, Newbury Park, CA: Sage.

Shakespeare, W. (1599) *Hamlet.*

The Holy Bible.

Willis, P. (2008) 'Getting a feel for the work: Mythopoetic pedagogy for adult educators through phenomenological evocation', in T. Leonard and P. Willis (eds) *Pedagogies of the Imagination: Mythopoetic Curriculum in Educational Practice*, Dordrecht, The Netherlands: Springer.

Willis, P. and Morrison, A. (2009) 'Introduction', in P. Willis, T. Leonard, T., A. Morrison and S. Hodge (eds) *Spirituality, Mythopoesis and Learning*, Brisbane: Post Pressed.

4

EPISTEMIC ASPECTS OF SOCIAL ECOLOGICAL CONFLICT

Richard Bawden

Prologue

I have personally been aware of *epistemes* and of the significance to human affairs, and of differences between them, for as long as my memory allows; of course I haven't always known them by that name. Over my lifetime, and until I was introduced to the work of the cognitive psychologist Karen Kitchener in the early 1980s (Kitchener 1983) I had progressively, and essentially non-reflexively, employed a whole lexicon of different alternatives for roughly the same notion. I probably first used nothing more insightful than 'opinion' to express the concept. After that, I would have progressed, in a pretty random and meandering sort of a way, through 'point-of-view', 'habit of mind', 'mental framework', 'perspective', 'position', 'frame-of-mind', 'filter', 'window-on-the world', 'view-of-the-world', 'worldview', 'prism', 'lens', 'mental-model', 'belief system', 'meaning-perspective', 'paradigm' and '*Weltanschauung*', before eventually encountering, appreciating and embracing the concept of *episteme*.

The central idea that I have come to accept in this regard is that through our experiences in the world, each of us comes to 'see' it (know it) from a particular epistemic position that reflects a set of assumptions about 'reality' to which we subscribe (essentially tacitly) and which finds expression in virtually everything that we do. Our assumptions about the nature of nature, about the nature of knowing and of knowledge, and about the nature of human nature, play themselves out in the way that we behave in the world about us.

It follows then that if we are to be more responsible in the way that we treat others along with the planet as a whole, we need to make sure that the epistemic positions to which we each subscribe, and those that characterise the respective cultures in which we are embedded, are adequate and appropriate to that task.

And to paraphrase Socrates, *first know thy episteme*!

Introduction

The character of our epistemes essentially shapes the way by which we create and maintain our relationships with others as well as with the rest of nature writ large. They are fundamental to the way that we each live out our lives and thus essential features of our social ecologies. This has huge implications with regard to the way that things are done as social ecological expressions of what are culturally accepted as desirable, feasible and morally defensible acts. Not the least of these implications is the fact that different people and different publics alike often hold to profoundly differing epistemes. And these distinctions can be the source of very significant tensions of conflict, which, not infrequently, tend to violence. These differences are manifest within and between all levels of social collectives, from families and small bands and clans through to entire cultures, nation states and indeed to whole civilisations. There is a profound mutuality here: cultures shape the epistemic assumptions of their individual members just as they, in turn, shape them. All cultures then are collective epistemic expressions of what is known and believed and assumed.

The situation is further complicated by the fact that each of us as an authentic individual has the capability of developing and changing our own particular epistemes over time and under changing circumstances. Furthermore, we are also capable of holding to different epistemic positions at one and the same time. Indeed it has been suggested that this ability to accommodate what we might call epistemic pluralism, is a sign of cognitive maturity (Churchman 1971). That said, we do seem to be incredibly and indefensibly reluctant to even challenge our own epistemes and those that characterise our own cultures, let alone change them even when and where there is a clear imperative for us so to do in the face of crises in our circumstances. While there is certainly a whole lot of evidence to suggest that an epistemic change is well and truly called for in these turbulent times, we seemingly remain immature in our epistemic development, and victims of our own epistemic intransigence.

The nature of our epistemes

To grasp the nature, and thence significance, of our epistemes more fully, it is useful to dwell a little on the idea of a 'worldview' that derives from the German word *Weltanschauung* meaning literally 'a perception of the world' (Funk 2001). Koltko-Rivera (2004), at the start of his comprehensive review of the psychology of worldviews, defines them simply as 'sets of beliefs and assumptions that describe reality'. At a later stage in that article he further elaborates this position arguing that worldviews embrace beliefs about what 'exists and what does not' (ontological assumptions), 'what objects or experiences are good or bad and what objectives and behaviours and relationships are desirable or understandable' (axiological assumptions) and 'what can be known … and how it can be known' (epistemological assumptions). To these three crucial dimensions, Funk (2001) adds four more: cosmological assumptions (beliefs about 'the origins and the nature of the universe, especially Man'), teleological assumptions (beliefs about 'the meaning and purpose of the universe, it's inanimate elements and its inhabitants'), theological assumptions (beliefs about 'the existence and nature of God'), and anthropological assumptions (beliefs about 'the nature and purpose of Man in general, and oneself in particular').

For the purposes here it will be enough to focus on ontology, axiology, cosmology and, most essentially, epistemology because it is fundamental to how we come to know about each of the others. It is these four intimately interconnected elements that I will use as the components of what I am referring to as our epistemes or our epistemic systems.

For me then, our epistemes represent the particular systems of valuing and values, knowing and knowledge, emotioning and emotions, believing and beliefs that we bring to bear to our everyday activities. It is our epistemes that shape the very way by which we continually transform our own personal and shared experiences of nature and of society alike into all of the different forms of knowledge that we use to inform the judgements that we make and the subsequent practical actions that we take. In this manner there is an inseparable interconnection between cognition and action: as Maturana and Varela (1987) indeed insist, 'all knowing is doing'.

Social ecology as epistemes-in-action

From this perspective, social ecology is both a formal knowing/doing endeavour as well as an integrated expression of 'ways of being'. This is captured well by Russell (1991) in his description of social ecology as 'a way of integrating the practice of science, the use of technology, and the expression of human values'. As he sees it, it is *social* because it reflects an underlying belief 'that it is people who make meaning', and it is *ecology* because its focus includes communities 'of the living and non-living things, and all the intricacies of their coherence and change'. Social ecology places significance on action, on critical reflection on that action, and on day-to-day living experiences, all within a context of both social and ecological responsibilities. Again to cite Russell, '[o]ne's acting in the world is seen to be the primary experience; how this experience is then interpreted and made sense of, flows from this essential experience as the actor reflects upon what happened' (Russell 1994).

In this manner, there are very significant distinctions between ecology and social ecology. In the first place, social ecology confronts head-on one of the epistemic difficulties that traditional ecology has faced with respect to the position of *Homo sapiens* within its field of studies. In formal, technical language, ecology is regarded as the scientific study of patterns of relationships between collectives of living organisms and their biophysical environments. This makes it, as May (1982) has asserted, 'a difficult science'. This is not just because, as he observes, 'evolution has given us only one world and it is not easy to perform experiments', but also because the subject matter – nature in the raw, as it were – is inherently complex. Born of the tradition of positivistic science with its realist ontology and objectivist epistemology, ecology seeks at the same time to escape from the conventional confines of reductionism and from the all too prevalent belief that there is only one way to do science. The paradox at the heart of ecology can be explained as follows:

> On the one hand, scientific biological explanations of the world have come to be seen by many as true explanations of how the world really exists. On the other hand, biological, ecological explanations of the world are understood to highlight the impossibility of humans ever knowing the world in an objective way.
>
> (Kuhn 1998)

Yet ecology remains firmly locked into a positivistic episteme that reflects an underlying objectivist belief that there is some 'permanent, ahistoric, matrix or framework to which we can ultimately appeal in determining the nature of rationality, knowledge, truth, reality, goodness or rightness' (Bernstein 1983).

The biologists Humberto Maturana and Francisco Varela present a somewhat different view. To these Chilean neurobiologists, we cannot talk of an independent 'external' world – *the* world – but only of *a* world brought forth in a 'languaging' between human subjects (Maturana and Varela 1987). Reality is thus a construction of social communication that reflects, in the process of construing, epistemic assumptions. Whenever and wherever these are in conflict they represent a state, as Maturana and Varela put it, of 'mutual negation' where the only way forward is to bring forth another world together: to create a different social ecology.

This situation, with respect to the 'doubtful' nature of reality, is further exacerbated by the problematic relationship that many ecologists recognise between *Homo sapiens* and the rest of 'nature'. As a species, we are either mere participants in the basic patterns of life on earth, and therefore subject to the great forces of biological evolution and to the seemingly endlessly changing bio–physical–chemical circumstances of the global environment, or we are beings who are somehow endowed by evolution with particular properties that allow us to 'step out' from that nature and somehow violate its natural patterns. It is from this latter perspective that it is claimed that the activities of *Homo sapiens* within nature are unnatural. It is through our own 'objective' observations that we are able to rationally state that many of our activities are destructive of nature.

So here we are, somewhat schizophrenically observing ourselves as participants experiencing nature who, by our own activities, are violating its 'naturalness' to the extent that we might even be representing its entire demise – even under circumstances where we should know (indeed do) better!

Berkes (1999) highlights the dilemma well in stating that within ecology, human beings are frequently regarded either as somehow 'un-natural' components of 'natural eco-systems' or are placed, somewhat poetically, into such mythological categories as the 'Ecologically Noble Savage', the 'Intruding Wastral', or the 'Fallen Angel'. However, a fourth category in this context helps us out of this bind and enables us to situate social ecology while also contributing to the significance to the episteme: *Homo sapiens* can be sensibly positioned as the 'Knowing/Valuing Being' (Bawden 2007). Through our cognitive competencies, we humans are capable (a) of coming to know about matters to hand that concern us because they invoke some value assumptions that we hold, (b) of coming to know how we come to know that, and (c) of coming to know the epistemic particulars that contextualise and 'shape' these two 'lower order' cognitive processes (Kitchener, 1983). Epistemic cognition thus concerns itself with reflection on 'the limits to knowledge, the certainty of knowledge, and the criteria for knowing' (Kitchener 1983).

If we are to bring forth these different worlds as a function of a quest to transform the way by which we live our collective lives in this age of modernity, we will need to modify the epistemes that come to dominate the modernist culture. If our interconnectivities and interrelationships with those in other cultures as well as with the rest of nature are to be developed in a manner that is sustainable, defensible, responsible and inclusive, we will need to establish epistemes that are appropriate to that task.

Epistemic transformation has become a moral imperative, and one of the earliest challenges in this regard is to subject the prevailing epistemic characteristics of modernism to critical review.

The prevailing episteme of modernism

For all the apparent success of our contemporary way of life in industrialised societies, and for all the improvements that have been legitimately claimed in the human condition through modern development, there is an accompanying, very significant downside. In pursuing the quest for progress through modernisation, we humans (unintentionally if not unforeseeably) have wreaked not inconsiderable havoc on both the biophysical and sociocultural environments in which we are embedded. The litany of these destructive impacts is as depressing as it is anxiety-provoking: from the global scale of the challenges posed by anthropogenic climate change, non-renewable resource depletion, and disastrous instabilities in financial systems and institutions, to the local extinction of fauna and flora, point source pollution, and the virtual extermination of the cultural belief systems of particular peoples. All are destabilising outcomes of well intentioned, if myopically focused human activities that have their foundations in the 'modernist' episteme.

The modern 'march of progress' has been accompanied by a host of consequences that, while unintended, have been destructive of both the biophysical and sociocultural environments that constitute our world. Such has been these impacts that some are now calling attention to the need to question the very foundations of modernism itself: as Ulrich Beck presents it, '[t]he transformation of the unseen side-effects of industrial production into global ecological trouble spots is … not at all a problem of the world surrounding us – not the so-called 'environmental problem' – but a far reaching institutional crisis of industrial society itself' (Beck 1992). And of course central to all of our cultural institutions are the foundational epistemic beliefs and assumptions that prevail. We are thus in the midst of an epistemic crisis of our own making – even if that is not appreciated by the citizenry at large in their 'epistemic ignorance'.

The genesis of my own personal commitment epistemic transformation can be traced to profound dilemmas that have confronted me as an agent of what might be termed the neo-agrarian epoch: not the agrarian age that was supplanted by the industrial age in the late eighteenth century, but the age that has seen the industrialisation of agriculture itself. On the one hand, as an agricultural scientist and educator, I have been among those who have contributed to the remarkable increases in food production. These, along with advances in medical science, have allowed the human population on the planet to increase from just over 2 billion at the time that I was born seven decades ago, to almost 7 billion as I write these words; a staggering statistic that indicates that we have well and truly escaped the ecological constraints that naturally shape the population dynamics of other species. Concurrently, if unintentionally, I have also contributed to the damage that our burgeoning population has imposed on both the ecologies and social ecologies of the world about us; with the processes of technological agricultural development themselves being prime suspects in that dynamic.

Somewhere along the line, however, as my life and career unfolded, I came to accept the folly of the unconditionality of the modernist view of technical development, whilst also recognising the inertia of the cultural episteme that supported it. I am of course far from being alone in that position. As Norgaard (1994) for instance has claimed, 'modernism betrayed progress by leading us into, preventing us from seeing, and keeping us from addressing, interwoven environmental, organizational, and cultural problems'.

Fundamental to the whole enterprise of modernity has been the almost universal uncritical acceptance of the logic and means of science and scientific inquiry without direct participation in the enterprise. Along with an acceptance, in good faith, of the benefits of science-based technological innovations, there has emerged an almost unconditional trust in scientists as the primary source of knowledge (although this does not seem to extend to situations where the focus is on the dis-benefits, as illustrated by the scepticism surrounding anthropogenic climate change!). What has developed through this exclusion has been referred to as a Culture of Technical Control (Yankelovich 1991) where, under circumstances that are in almost direct contradiction to the modernist appeal to democracy, 'expert thinking in science, technology, economic enterprises, government, the policy sciences, and large organizational structures' has come to control (dictate) the way that things are done: the manner by which lives are lived.

Yankelovich's position reinforces the sombre view of Toffler (1984) that the 'political technology' that has emerged as a consequence has not been particularly adaptive. Indeed, as he put it, a mismatch has arisen between 'our decisional technology and the decisional environment' that has been characterised by 'a cacophonous confusion, countless self-canceling decisions, noise, fury, and gross ineptitude'.

It is not that the linearity, objectivism and reductionism of techno-science and neoliberal economics are 'wrong', for they have indisputably brought prosperity and improved well-being to countless millions. By and large, at least when evaluated from a utilitarian perspective that relies on benefit/cost ratios for its justification, modernism has been an eminently successful episteme. However, there are now significant signs that it is an inadequate and overtly inflexible episteme in the face of the complexities of everyday living, and as an expression of the values that citizens espouse in their everyday lives that extend beyond an acceptance of ends justifying the means by which they are attained. Witness the widespread public confusion about the nature of nature and the angst about how it should be treated. If land has no sacred dimension in this modernist world of ours, but is merely taken as a resource to be managed with its landscapes shaped to meet human aesthetic values, then how is it now to be handled without further damage being done to its integrity? And if *Homo sapiens* is no longer regarded as an integral component part of nature and subject to its ecological 'laws', how are people to behave with respect to their relationships with other species as well as with the land itself?

There surely must be lessons that can be learned from exploring other cultures where shades of pre-modernism can still be detected – and indeed where the full impacts of the modernist hegemony can still be revealed in alternative narratives. And a prime example of such a narrative is that provided by Bob Randall, an elder of the Anangan peoples of Central Australia in a 2006 documentary – *Kanyini* – which was produced by Melanie Hogan (Documentary Australia Foundation 2006).

The Kanyini narrative

In the documentary, Bob Randall submits that the overall (epistemic) commitment of his peoples is to 'responsible connectedness' – *Kanyini* in the Anangan language. This finds expression in what he describes as the four pillars of that commitment. Each of these elements can be seen to have equivalence to the essential features of epistemes as they have been recognised among modern psychologists and philosophers. *Tjukurrpa*, translated by Randall as the overarching system of beliefs, can be seen to constitute the overall Anangan episteme as it represents the 'Dreaming' and provides the foundations of *Kanyini. Ngura* – land as nature and home alike – can be considered to represent an ontological position, while *Kurunpa* – pantheistic spirituality and the sanctity of the ancestors – presents a metaphysical cosmological perspective. Finally *Walytja* – family, in its broadest possible connotation – might be seen as that which has a special value and morality with respect to human nature, and thus an axiology.

Kanyini, the documentary, is not merely an account of the epistemic foundations of the Anangan peoples of Central Australia, but, as Randall forcefully narrates, a searing indictment of the destruction of each of the four epistemic 'pillars' in turn, at the hands of non-indigenous settlers and their successors. The essential outcome of this epistemic assault has been, as he submits, the devastation of an entire culture and the decimation of a once self-fulfilled self-worthy society. In its remnant state it now characterises the zero-sum nature of the clash of epistemic cultures – the asymmetrical conflict between two fundamentally different social ecologies.

The introduction and subsequent pioneering development of agriculture to the Australian continent by the European settlers in the late eighteenth century assuredly did represent an epistemic assault to the essences of both *Ngura* and *Karungpa*: offences against beliefs both about the nature of the relationships of humans with the land, and about the spiritual basis of those beliefs.

First came the introduction of exotic domesticated animals onto a landscape that had never to that point experienced cloven-hoofed ruminants (or indeed any domesticated livestock for that matter). The introduction of fences, dams, water bores, windmills, machinery and a range of constructed infrastructures that followed was not only a further affront to the Aboriginal episteme, but a series of physical barriers to their traditional hunting practices – and indeed to their essentially nomadic way of life as gatherers as well as hunters. Cultivation for cropping further greatly exacerbated this provocation to their social ecology.

But perhaps most significant of all with respect to epistemic differences and attendant social ecological tension was the imposition of laws relating to ownership of the land by the settlers. This denied any legal rights to land to those Aboriginal people who were, without any doubt whatsoever, its traditional owners – or, from their own perspectives, responsible and empathic stewards of a nature with which they were inextricably interconnected and in which they were firmly embedded as component parts of an indivisible whole. From this perspective, the introduction of agriculture in Australia was also a severe epistemic offense to the 'pillar' of *Walytja*, for nature was 'family' to Aboriginal people. The process of land clearance was thus tantamount to fratricide. Of even greater import here was the actual destruction of the fabric of human families, and the very autonomy of an entire peoples, through the adoption of a process known as absorption. In essence, starting in the mid-nineteenth century, the by-then dominant culture of the non-indigenous people expressed in action the epistemic belief in the 'rightness' of their rights to 'take-over' Aboriginal Australians and determine the way that they should live their lives. In many instances, this represented the creation of a second-class citizenry who lived their lives by courtesy of the patronage of others. It led to instances of human rights abuses, cruel exploitation and, most disgracefully, indefensibly and tragically of all, to killings.

It is true that agriculture gained a whole new workforce where, contrary to the expectations and prejudices of those non-indigenous people who equated 'ancient' with 'primitive', a great majority performed with distinction. Importantly it also illustrated that with respect to certain elements at least, 'traditional' epistemes were flexible. Aboriginals could live and work with non-aboriginals without rejecting their overall belief systems even while, until the mid-1960s, their representative rights were ignored. Almost three decades further were to pass before the High Court of Australia handed down a judgement that essentially recognised the rights of Aboriginal peoples, under certain prescribed circumstances, to hold legal title to land. This decision was characterised by a very unusual epistemic sensitivity by the Court that was expressed in the explicit recognition that the relationships between Aboriginal people and the land was marked more by spiritual dimensions than material ones (Wooten 1995).

The third epistemic assault came in the arena of *Kurunpa* when the monotheism of the Christian settlers confronted the pantheistic spirituality of the Aboriginal peoples. This once again had the essence of a zero-sum game, with one cosmology entirely dominating the other, ironically in the name of God! There are fewer better arenas for exploring epistemic distinction than metaphysical cosmologies, for there are so many differing intercultural notions of what constitutes the origins of the universe from the secular to the sacred. Disputes also continue to reign about the nature of spirituality or, indeed, of religion. All of this notwithstanding, there is incontrovertible evidence of the central significance of spiritual beliefs to the social ecology of the traditional ways of life of the Aboriginal people of Australia. As Flood (2006) has observed, 'Aborigines see traditional society as based on the spiritual', as it relates both to the genesis of the universe and the sanctity of the spirits of the Ancestral Beings.

The greatest casualty in this clash of epistemic cultures would find its expression in the apparent complete lack of comprehension by the non-aboriginal settlers of the overarching system of beliefs of Aboriginal peoples as expressed through the 'Dreamtime': the *Tjukurrpa* of the Anangan peoples. Among the Australian Aboriginals, the 'Dreamtime' or, in conveying its timelessness, the 'Dreaming' (Stanner 1953), is an absolutely central epistemic feature of their culture. It is this 'complex network of faith, knowledge, and ritual' that dominates 'all spiritual and practical aspects of their life' (Flood 2006). From this it might be concluded that this is at once the epistemological basis for the way that at least some Aboriginal people came to know about the world about them, the ontological basis for their beliefs about that world and their place in it, the axiological basis for how they should live their lives in it, and the cosmological basis for the sanctity of it all.

Lessons from Kanyini

The four Anangan epistemic distinctions provide a highly relevant framework both for raising epistemic awareness and then investigating the potential for exploiting synergies between different epistemes, as the sources of social ecological transformations in order to 'bring forth new worlds together' – to the mutual benefit of both peoples through the synergistic opportunities presented by epistemic pluralism.

From within a perspective of interconnectedness (*Kanyini*), we could/should seek a critical review of our overall epistemes (the spirit of *Tjukurrpa*) and explore potential synergies between holo-centrism and techno-centrism for a post-modern world. The reunification of values with knowledge would lead to a new epistemology, while the contextualisation reductionism with holism would point to the need for a new ontology. We could/should revisit different ontological views on the nature of nature (the spirit of *Ngura*) and create a new land ethic and a better philosophy for the use of finite natural resource and the management of the national estates. We could/should seek to explore the whole cosmological spectrum of human spirituality (the spirit of *Kurunpa*) in a search for a post-modern morality that better suits us for dealing with the wide range of ethical dilemmas that modernism has revealed – from religious fundamentalism through to the 'rights of nature'. And we could/should examine the vast issue of the nature of human relationships (the spirit of *Walytja*) across the entire range of human organisation from family, through communities, through governments, through transnational corporations, through educational institutions, to entire nation states and beyond, to the unity of the human race itself.

We need a social ecology of critical engagement marked by a commitment to systemic wholeness through participation:

> Wholeness means that all parts belong together, and that means that they partake in each other. Thus from the central idea that all is connected, that each is a part of the whole, comes the idea that each participates in the whole. Thus participation is an implicit aspect of wholeness.
>
> (Skowlimowski, 1985)

Conclusion

This is not a call to return to the traditional 'ancient' ways of living reportedly enjoyed by at least one clan of Australian Aboriginals pre-European settlement. That would neither be desirable nor possible – even for the Aboriginal peoples themselves. Nor is this a claim for the total rejection of the 'modern', even where some of the most undesirable consequences of the industrialised way of life have been emphasised. Rather it is to point out that essential differences between 'ancient' and 'modern' cultures as revealed through Bob Randall's *Kanyini* narrative are reflections of profoundly different sets of epistemic beliefs and assumptions. These epistemic differences have assuredly led to high levels of conflict over the past two hundred years or so that have tended to play out as a zero-sum game between two social ecologies: there has been a winner and a loser, as one epistemic culture has come to essentially overwhelm the other. One coherent episteme has been under severe challenge in the face of the 'march of the progress' of the other. Yet each could have learned so much from the other, for each has particular beliefs with respect to the way that lives *can be* and *ought to* be lived that are as wise as they are responsible and that could lead to both resilience and sustainability.

There are vital aspects of Aboriginal holo-centricity that surely would provide epistemic contexts for amendments to techno-centricity that would ameliorate some of the negative consequences that continue to flow from its expression. The reunification of knowledge with values and emotions would be one such advantage. People could be whole again. A closer alignment of people with their social and ecological environments would be another. People could be truly communal again and at one with their natural world. These are not romantic ideals. It has been their absence that has been the worst of the negative impacts of modern development – not just within Australia, but across the whole of the 'developed' world. The techno-centric foundations of the process of modern development itself are proving to be inadequate in the face of so much deterioration in both biophysical and sociocultural environments across the globe.

Epilogue

I am not unfamiliar with the challenges provided by epistemic transformation for, one way or another, I have spent an entire professional career exploring the role of higher education in promoting precisely that – essentially within a context of the search for a 'better' episteme for agricultural and rural development (Bawden 2005). It has been a far from easy 'road less travelled' yet the journey has been one of immense personal fulfilment along with the occasional socioecological achievement.

It is from the experience of such fulfilment that I draw inspiration for continuing my efforts – my three score and ten years notwithstanding. It is also the source of my unquenchable optimism that the world can and, with commitment to epistemic transformation, will be a better place.

References

Bawden, R.J. (2005) 'Systemic development at Hawkesbury: Some personal lessons from experience', *Systems Research and Behavioural Science*, 22: 151–64.

Bawden, R.J. (2007) 'Knowing systems and the environment', in J. Pretty, A. Ball, T. Benton, J. Guivant, D. Lee, D. Orr, M. Pfeffer and H. Ward (eds) *Sage Handbook on Environment and Society*, London: Sage Publications, Chapter 15, 224–34.

Beck, U. (1992) *Risk Society: Towards a New Modernity*, London: Sage Publications.

Berkes, F. (1999) *Sacred Ecology: Traditional Ecological Knowledge and Resource Management*, Philadelphia, PA: Taylor & Francis.

Bernstein, R.J. (1983) *Beyond Objectivism and Relativism*, London: Blackwell.

Churchman, C.W (1971) *The Design of Inquiring Systems*, New York: Basic Books.

Documentary Australia Foundation (2006) http://www.documentaryaustralia.com.au/da/caseStudies/details.php?recordID=30, accessed 8 February 2011.

Flood, J. (2006) *The Original Australians: Story of the Aboriginal People*, Sydney, NSW: Allen & Unwin.

Funk, K. (2001) 'What is a worldview?', http://web.engr.oregonstate.edu-funkk/Personal/worldview.html, accessed 10 February 2010.

Kitchener, K. S. (1983) 'Cognition, metacognition, and epistemic cognition: A three level model of cognitive processing', *Human Development*, 26: 222–32.

Koltko-Rivera (2004) 'The psychology of worldviews', *Review of General Psychology*, 8: 3–58.

Kuhn, L. (1998) *An Investigation of Mystery in Ecological Knowing*, unpublished PhD dissertation, Melbourne, VIC: Deakin University.

Maturana, H. and Varela, F. (1987) *The Tree of Knowledge: The Biological Roots of Human Understanding*, Boston, MA: New Science Library, Shambahla Publications.

May, R.M. (1982) *An Overview: Real and Apparent Patterns in Community Structure*, in D.R. Strong, D. Simberloff, L.G. Abele and A.B. Thistle (eds) *Ecological Communities: Conceptual Issues and the Evidence*, Princeton, NJ: Princeton University Press, Chapter 1.

Norgaard, R.B. (1994) *Development Betrayed: The End of Progress and a Co-evolutionary Revisioning of the Future*, London: Routledge.

Russell, D.B. (1991) 'Social ecology in action: Its rationale and scope in education and research', *Studies in Continuing Education*, 13: 126–38.

Russell, D. B. (1994) 'Social ecology – education and research', in L. Fell, D.B Russell and A. Stewart (eds) *Seized by Agreement, Swamped by Understanding*, University of Western Sydney: Hawkesbury Printing.

Skowlimowski, H. (1985) 'The co-operative mind as a partner of the creative evolution', *Proceedings of the First International Conference on the Mind–Matter Interaction*, Brazil: Universidad Estadual de Campinas.

Stanner, W.E.H. (1953) *The Dreaming*, in W.E.H. Stanner (1979) *White Man Got No Dreaming: Essays 1938–1973*, Canberra, ACT: ANU Press.

Toffler, A. (1984) *Introduction on Future-Conscious Politics*, in C. Bezold (ed.) *Anticipatory Democracy: People in the Politics of the Future*, New York: Vintage Books.

Wooten, H. (1995) 'Mabo and the lawyers', *Australian Journal of Anthropology*, 6: 116–33.

Yankelovich, D. (1991) *Coming to Public Judgment: Making Democracy Work in a Complex World*, Syracuse, NY: Syracuse University Press.

5

CHAOS, REACTION, TRANSFORMATION

Bernie Neville

> Every few hundred years in Western history there occurs a sharp transformation. Within a few decades, society – its worldviews, political structure, its arts, its key institutions – rearranges itself. Fifty years later there is a new world … We are currently living through such a transformation.[1]

At the beginning of the twentieth century reasonably aware people in Australia had no doubt that science and technology were making the world a better place. Medicine was well on the way to eradicating disease. Slavery had been abolished forever. Australia and the rest of the world were inevitably becoming more educated, more prosperous, more contented. Yet at the beginning of the twenty-first century we have a dozen reasons to conclude that human society, indeed the planet, is on the edge of destruction: environmental degradation, overpopulation, a deepening abyss between the rich and poor, the exhaustion of non-renewable sources of energy, the greenhouse effect, holes in the ozone layer, the starvation of millions, global warming, melting ice caps, rising sea levels, disappearing biodiversity, tribal conflict, uncontrollable disease, an ongoing and increasingly destructive arms race, the global slave trade, the displacement of populations, the global drug trade, pre-emptive wars, the engineering of hate and fear, epidemic mental illness, human alienation and frustration on a vast scale.

The sense of impending disaster is not new. We have been living with it for half a century and we have found ways of coping with it psychologically.

In the aftermath of the Second World War, Jean Gebser was arguing that 'the crisis of the world is in a process … of complete transformation, and appears headed for an event which, in our view, can only be described as "global catastrophe"'.[2] He went on to suggest that 'only a few decades separate us from that event' on the grounds that the increase in technical feasibility appears to be inversely proportional to our sense of responsibility. We seem to be doomed, 'unless a new factor were to emerge'.

Gebser was prepared to assert that this 'new factor' that might enable us to survive the crisis was in fact emerging in the mid-twentieth century. He accumulated evidence from the arts and sciences to back his assertion. He argued in *The Ever Present Origin* that, just as the crisis is 'not a crisis of morals, economics, ideologies, politics or religion', neither is the new factor to be described in any of these ways. The crisis is a crisis of consciousness, and the 'emerging possibility' is a transformation of consciousness.

When Gebser was writing about the planetary emergency, he had in mind the Cold War and the prospect of Mutually Assured Destruction. For decades we feared the imminent nuclear winter. We took a moment to draw breath at the beginning of the nineties and then fell once again into deep anxiety, this time about the imminence of global warming. Global warming had been going on since the beginning of the industrial age, and we've no guarantee that the nuclear winter has now been avoided. We have plenty of other things to worry about now but, whatever it is that we choose to worry about, we have evidence enough that we are currently in the midst of an era of ecological crisis and psychological incoherence, and our leaders – global and national – seem to be unable to address the crisis effectively.

One aspect of this psychological incoherence is our unwillingness to see the deep connections between the environmental crisis and worldwide social and political disruption. There have presumably always been poor people on the planet, yet in the twenty-first century there are millions who are totally destitute. According to the World Health Organization, there are at least 500 million people with serious mental illness or addiction; over twenty million people attempt suicide each year, and many of them succeed.[3] Serious mental distress among adolescents appears to be as prevalent in Australia as it is in Europe and North America. We find violence against humanity in every part of the globe, leaving the millions who survive it suffering from post-traumatic stress. In a complex, dynamic, organic system, such as the one of which we are part, all these elements – including the handful of refugees who come in leaky boats to Australia – are interconnected.

The paradox in all of this is that, though we are well aware of these dangers, we do almost nothing to confront them. We have known for decades that the resources of the planet are running out, yet this information has minimal impact on behaviour. We do little in our day-to-day lives to address the crisis, and political attempts to confront it at a global level meet with uncompromising resistance. Indeed, our political masters keep expressing their delight at rising production and consumption, when rising production and consumption are what is killing us. The stresses on our planet and our personal and collective pathology are deeply connected.

Why are we so unaffected by the horrors that lie in wait for our children? Why do we do nothing, or at least so little that it will make no difference?

The political response to crisis in Australia appears to be largely a reactive one. Confusion is simplified and national identity is sustained by collectively employing our psychological defences: denial, projection, fixation, displacement, dissociation, splitting and psychic numbing. These defences may enable us to impose some form of order on our experience, they may help us to avoid being traumatised by the horror of our situation, but they only make the crisis worse. When we experience the anxiety aroused by an uncertain future, we are as likely to engage in 'retail therapy' as in environmental activism. A rational response to crisis would be to assess the risks to the planet and reflect on how we could best address them. Instead, we find national leaders publically unaffected by the conclusions of climate scientists, and a liberal democracy compliant in their denial.

The apologists for reactive defences against the planetary crisis argue that the only alternative to chaos is certainty. They can point to many places on the planet where chaos rules, and in comparison to the genocides, terrorism, corruption, avoidable famine and pestilence, and the anarchic criminality we see elsewhere, a consensual, highly regulated Australia certainly seems attractive. The message that we must defend ourselves against all these terrors by reinforcing our geographical and psychological boundaries finds a receptive audience.

We may speculate that it is our deep collective anxiety about an uncertain future that is behind the move towards increased regulation in Australian education. Where we might expect that education systems would look to the future and become key sites for a global revolution of consciousness, we find increasing levels of government control over Australian schools – not with a view to encouraging initiative, innovation, diversity and creativity – but to ossify the ordinariness that currently characterises it. Australian ministers of education, both state and federal, have formalised 'evidence-based' values through the implementation of competence standards in teacher education, aiming to convert teachers and school principals from professionals to 'knowledge workers' and 'knowledge managers' who will carry out the instructions of their superiors. Competence is increasingly defined in terms of the ability to comply with externally applied edicts. The reactive response to the threat of uncertainty is manifested in 'back to basics', a nationally imposed curriculum, measurable outcomes, standardisation, regulation, central control, test-driven teaching and constant assessment of whatever is assessable.

Governments pushing the reform of education are inclined to argue that these measures are a unique national response to unique national conditions, that they are designed to achieve better educational results for the majority of students and that they are designed to increase access and equity in education. Such reforms are perhaps better seen as a panicked reaction to a world that is changing in ways that national leaders cannot control. In the Australian context there is a great deal of rhetoric about a commitment to quality in education, and the importance of an educated and skilled workforce and the significance of education as an export industry, yet both public schools and state universities are being starved of resources.

Rather than confront the planetary crisis, which would demand a population educated to understand the crisis and their part in it, rather than promote a high quality education as a product for the Asian market, which currently manifests a high demand for it, and rather than label money spent on education as an investment instead of a cost, the national government devotes considerable energy to persuading young people to abandon the idea of a genuine university education. Educational objectives have been restated in terms of labour market needs, in response to the demands of global markets. Yet education itself is one of the country's top three products in terms of export income. To allow the quality of this product to deteriorate through lack of adequate funding, as is currently happening, is more readily explained as panic in the face of imagined chaos than as a rational response to a carefully analysed situation.

At a time when information technology and ease of travel are making global community possible, the response of the Australian government to the evidence of dissolving national and cultural boundaries manifests the same signs of panic. Where once we had a rhetoric of diversity and were proud of the way our schools respected and supported the values and cultures of the many immigrant children who populated them, we increasingly find a pressure to conformity. The unlamented Howard government, under the leadership of a prime minister who found people from different cultures wearing traditional dress 'confronting', ensured that Australia ceased to be proclaimed in federal government documents as a multicultural society, and that the promulgation of 'Australian values' came to be seen as a matter of urgency.

Gilles Deleuze[4] associated the common human need for cohesion, stability and autonomy with the 'end of enclosure'. He argued that sites of enclosure are failing and that attempts to maintain traditional boundaries are accompanied by an ongoing condition of uncertainty and anxiety, which in turn leads to a hardened resolve to maintain one's boundaries and one's identity. Even though Australia's prosperity has been built on immigration, the inhumane treatment of attempted refugees over the past couple of decades provides another indicator of panicked reaction to crisis. Open our borders to the oppressed and the destitute and we will have chaos. Close the borders around the tribe and we will be certain of who we are and we will be safe.

The boundaries that once surrounded the educational enterprise are also under attack. Where schools were once the sites where key educational understandings were first encountered and engaged with, these encounters and engagements are increasingly being experienced outside of them. Educational television and the Internet provide attractive alternatives to schooling as sources of knowledge. This is a challenge for those whose authority rests with their expertise as knowers. It is not surprising to find teachers reacting defensively to this threat to their identity. Though many teachers may perceive their own thinking and values to be at odds with the demands of their political masters, we should not be surprised to find them behaving in a way that is psychologically similar to that of the ideologues who want to regulate them. They may be environmentally conscious and may castigate the government for its paralysis in the face of social and environmental crisis, but if they cling to entrenched notions of their role, remain closed to the possibility of 'consciousness transformation' and fail to embrace the global connectivity enabled by the information and communications revolution, they may be resisting a critically important means of dealing with this crisis.

In October 1989 Vaclav Havel was arrested by the most conservative communist regime in Eastern Europe and sentenced to two months jail. Five months later he was making a speech to the US Congress as president of an emerging democracy. Such things are clearly possible. However, he was well aware that:

> Without a global revolution in the sphere of human consciousness, nothing will change for the better in the sphere of our being as humans, and the catastrophe toward which this world is headed – be it ecological, social, demographic or a general breakdown of civilization – will be unavoidable.[5]

The situation of the planet has clearly deteriorated since 1990, and we are still in denial, even those of us who can talk the talk about environmental emergency, environmental responsibility and environmental agency. We are still inclined to regard the environment as 'other' – something we have, something we live in, something we may choose to exploit or choose to care for. We are still inclined to imagine the starving and traumatised people of Darfur as 'other', and even more likely to categorise their oppressors in the same way. We even distinguish between the 'good guys', like ourselves who are sensitive to social and environmental distress, and various 'bad guys' (the transnational corporations? the Americans? the Japanese whalers? the pharmaceutical companies? the oil sheiks? the right-wing think-tanks?), who are responsible for it. It would be better to acknowledge that we ourselves are the 'bad guys' who are endangering the planet and its people. As long as we project responsibility for environmental and social catastrophe onto others, and as long as we fail to identify with the planet and all its human and non-human inhabitants, we are trapped in the mindset that is responsible for the imminent catastrophe.

Most Australians are still trapped in the myth that 'everything will turn out all right in the long run', that some one, somewhere, will be clever enough to invent the technical fix that will solve the problem. Most are still, as Havel argues, 'under the sway of the vain belief that man is the pinnacle of creation and not just a part of it', failing to acknowledge that 'the anonymous mega-machinery we have created for ourselves no longer serves us but rather has enslaved us'.[6] To think otherwise requires a substantial transformation of consciousness.

It is nearly seventy years since Gebser set about the massive accumulation of evidence that persuaded him that a new cultural pattern was emerging. In the new century we find, both in the arts and sciences and in popular culture, abundant evidence that this pattern has intensified. And we find a readiness in both arts and sciences to conclude that the rational, logical, sense-based thinking that made modern scientific, industrial culture possible no longer seems to be adequate. Our sense of self has become conditional, situational and transitory. The old, dualistic distinctions between truth and error, male and female, human and non-human, subject and object, matter and spirit, image and reality, self and other, good and evil no longer provide the essential scaffolding for our thinking, unless we are still finding security in the conventional fundamentalisms of our inherited globalising, colonising tribal consciousness.

The end of enclosure may bring the many negative consequences of globalisation, but it also brings the promise of globalism – the belief that the well-being of every person, no matter how far away, affects us all, and that we share a fragile planet that will not survive unless we learn to respect all beings, human and non-human. The fate of the planet will not be determined by the brilliance of our technology, but by the genuineness of our dialogue and the strength of our connection to all life. Globalism represents a significant transformation in the way human beings think about the world and relate to each other. A shift away from individualism, tribalism and a narrow scientific–technological materialism to a sense of the species, the planet and the cosmos as an organic whole, may be seen as evidence of a revolution of consciousness, a mutation from a collapsing rational structure to Gebser's 'new factor'.

If there is no revolution of consciousness, we appear to be in trouble for, to quote Havel again, 'the salvation of the human world lies nowhere else than in the human heart'.[7] However, in an optimistic moment we may reflect on the possibility that something is stirring in human consciousness, and that our students will be more at home in what Gebser calls 'the integral structure of consciousness' than their teachers are. They may be more comfortable with complexity and ambiguity. They may be able to hold competing 'truths' in tension rather than insist that truth is single and universal. They may more readily feel empathy for all beings, human and non-human, than their parents do. They may be less convinced that the unspoken assumptions of their national or ethnic culture represent a universal truth. They may more readily identify with 'the whole' rather than with the fragment of life confined within their skins. Their easy acceptance of information and communications technology may give their personal decisions a global resonance that was not available to their parents.

In his cautious review of climate change commissioned by Australian federal and state governments, Ross Garnaut made it clear that denial and defensive reaction are no longer options. Not to act immediately in cooperation with developed and developing nations is to ensure catastrophe. A 'business as usual' approach will result in a 5.6 degree increase in global temperatures in the course of the century, resulting in 'collapse and chaos' that would destabilise virtually every aspect of modern life.[8] Garnaut has no doubt that we face 'a diabolical problem' in developing domestic policies consistent with ultimate international agreement. However, he is encouraged by the evidence of a transformation of national consciousness manifested in community support for action on climate change, 'even if it will cost them and if Australia has to act unilaterally'.

It is hardly necessary to emphasise the role that teachers play in this. Regardless of their control over the curriculum, or their lack of such control, it is clear enough that teachers model attitudes and actions that have substantial impact on the attitudes and actions of their pupils. If it is indeed true that we are witnessing in ourselves and our students an emerging consciousness that has the capacity to face the present planetary emergency without denial or defensive reaction, and if we know how to make use of the new possibilities for human connection across nations and across cultures, our local and apparently insignificant contributions may make a difference. As Garnaut observes in concluding his report, 'There's just a chance.'[9]

Notes

1 Peter Drucker (1992) 'The post-capitalist world', *Public Interest*, 109: 89–101.
2 Jean Gebser (1985/1949) *The Ever-Present Origin*, Athens, OH: Ohio University Press, pxxvii.
3 World Health Organization (2001) *World Health Report 2001: New Understanding, New Hope*, www.who.int/whr/2001, accessed 4 February 2011.
4 Gilles Deleuze, (1992) 'Postscript on the societies of control', *October*, 59: 3–7.
5 Vaclav Havel, Address to US Congress, 21 February 1990.
6 Ibid.
7 Ibid.
8 Garnaut, R. (2008) 'Climate change review: Targets and trajectories', address to the National Press Club, Canberra, 5 September.
9 Ibid.

6

THE BURDEN OF NORMALITY AND THE PROSPECT OF MORAL IMAGINATION

Barry Bignell

'Be not deceived when Death pretends to offer life!', utters Persephone, Hades' spouse, to Orpheus at a decisive moment in Owen Barfield's play of that name (1983: 72). Hades, Barfield informs us, is the tyrannical god for whom the number two is sacred, and is also 'the region where the principle of lifeless repetition has triumphed' (1983: 116). Mindless repetition, the comforting illusion of habit is represented in the play by Sisyphus, a tormented soul in Hades who is compelled to push a huge boulder to the top of a hill, only to have it roll back to the bottom at every attempt. Since Sisyphus has lost all memory of his former human self, it has not occurred to him to question the absurdity of his action, for he is grateful to have a job. Albert Camus, drawn also to the Sisyphus myth, gave the term 'the absurd' to 'that unspeakable penalty in which the whole being is exerted toward accomplishing nothing' (1991: 120). In the tragic character of Sisyphus we perceive a modern human predicament. 'Man alone', Barfield comments on the play, 'can deliberately will the *repetition* of an experience. And repetition, experienced as such, is at the heart, for good and evil, of his faculty of reasoning, and thus makes possible his language, his art, his morality, and indeed his humanity. Yet it is the enemy of life, for repetition is itself the principle, not of life but of mechanism' (1983: 116). Persephone's warning, then, is an exhortation to be aware of the life-denying impulse of habit masquerading as purpose, no matter how 'normal' repeated experience may seem. For my purposes here, I will call two-ness, the same again, automatism, marking time, sterile effort, epistemic inertia and the familiar 'the burden of normality'; burden because normality blinds us to the possibility of raising experience to consciousness.

It is that blindness and that possibility, the subjunctive mood, the 'what if?' with which this chapter is concerned. Why does life for many seem to be a matter of menu options, a purposeless drifting from one fad to another? Why do we so readily forget that being human is an open project? Why, in our knowing, do we seem to be prisoners of presentism, 'the assumption that the time, place and culture within which we happen presently to live somehow represent the limits and summation of human experience and wisdom' (Donaldson 2009: 40)? Why, in the face of all the evidence, do we accept far less of ourselves than we know we are capable of?

I came across a *provisional* solution to the problem of purposelessness some twenty years ago while reading Rudolf Steiner's seminal text, *The Philosophy of Spiritual Activity* (1992), in which I first encountered the term 'moral imagination'. Steiner's view was that we live on a far lower plane of value than the epithet 'human' merits, and that a significant vision of human existence must reach beyond biological explanation to include the psycho-spiritual dynamics by which we are, *as* humans, inalienably implicated in the world process. Almost unknown to us today, moral imagination is the power to envisage ourselves as more than we are, of being conscious of our humanity in all things. I say provisional because the adjunct 'moral' implies a choice. A choice always involves some kind of resistance, or it is not a choice. One must inevitably choose *for* something and *against* something else by becoming aware of alternatives. Provisional, moreover, because imagination as it is meant here is a systematically developed facet of intelligence, distinguished from fantasy and spontaneous flights of fancy. Moral imagination is a kind of epistemic competence generated by the individual will. Since it can only be freely chosen, it can be neither inculcated nor enforced. It is important, then, not to confuse legislated freedoms – freedom of speech, of worship, of assembly, of expression and so on – which have been handed to us, with freedom of consciousness, which is *achieved* through personal effort. Purpose or chance? Consciousness or mechanism? The choice is ours. Psychologist Franz Winkler articulated the idea of moral imagination concisely: 'Being free, man can remake himself in the image which he forms of himself, and the moral choice of our time will be reflected by the likeness this image will assume' (1960: 167–8).

I propose, given limited space, to subject only one example of human behaviour, namely, speech, to imaginative re-creation. I hope to demonstrate that the familiar concept of intelligible utterance can be expanded to include a deeper understanding of the human being.

There has been much concern expressed in the media recently about declining levels of literacy in Australian schools, and heated debate about the methods by which children best become literate. Children, government authorities insist, must learn to read and write. The 'logic' of this imperative seems to rule out further deliberation, perhaps because of the tacit recognition that lack of facility in the use of words severely inhibits a person's progress in the world. In failing to distinguish between the spoken and printed word, however, it risks reducing language in total to a utilitarian key of communication. It veils the *talent* for speech and the question of what it means to be a being in possession of language in the context of human existence as a whole. A question guided by moral imagination, then, might be formulated thus: 'What particular psychological image or guiding profile of the child and childhood do those who formulate public policy have in mind when they speak about the literate person?' It is no accident of etymology that *literate*, 'able to read and write' and *literal*, 'to the letter,' both derive from the Latin, *littera*, letter. Both connote the absence of allegory and metaphor, that is, imagination. If literacy is about matching visible signs, letters, to sounds, and combinations of letters to words, then it is a conversion of something belonging to aural experience into visible form. It is, in short, an abstract operation.

In giving primacy to literacy as we usually think of it, are we not exposing children to abstraction before they are ready for it? Are we not inadvertently projecting an image of our adult selves onto them? If so, why are we in such haste to turn children into miniature replicas of ourselves? Can we honestly rule out social usary, the ideology that sees children increasing in value the more literate we make them, the earlier the better, so that they do not become a 'burden on society?' Does it not make all the difference to the potency of a word or a sentence whether there is a person, rather than a human thing, behind it? Children do not learn by abstraction (which comes later); they learn by experience. First, there has to be the *experience* of speech.

What comes to mind when you hear or see (for you read as much with the ear as with the eye) the term, *Ayers Rock*? Something must come to mind, or *Ayers Rock* would be a mere noise and not words. This coming-to-mind involves your bringing up from within a concept to meet what you *hear* when I say or when you read *Ayers Rock*. The fusing of the concept, rock, and the sound you hear is what converts the noise into a name. To qualify it with the possessive Ayers makes no difference to its status as rock.

It is easy to understand that this naming process is habitual because we are not conscious of it happening. It is *not* so easy to understand – because it is normal – the extent to which it guides and even confines our knowing and limits our capacity for (what we loosely term) experience. Michael Polanyi explains:

> As observers or manipulators of experience we are guided *by* experience and *through* experience without experiencing it *in itself*. The conceptual framework by which we observe and manipulate things being present as a screen between ourselves and these things, their sight and sound, and the smell and touch of them transpire but tenuously through this screen which keeps us aloof from them.
>
> (1969: 197)

We in the present epoch stand outside our phenomena. It has not always been like this; nor does it have to continue. At this point I invite the reader to join me in an exercise. Sound the vowel *u* (as in *oo*) to yourself! Repeat it! Try singing or intoning it! Give it wings! Notice that you can sustain it for as long as your breath lasts! That is why vowels, not consonants, form the substance of song. Notice how the lips extend outwards from the body (when compared to the vowel *a*, as in mask)! The sound leads, or points. Now place an *l* before the *u* so that you sound the syllable, *loo*! Play with it! Taste it! Notice what your tongue does in order to sound *l*, and how fluid *oo* becomes when you join the *l* to it! We might say that these two sounds *belong* together. This belonging is encapsulated in the other meaning of the verb articulate, 'to couple together'. Now place an *r* before the *u*, so that you sound *roo* (the *r* can be rolled or not!). Again, notice the movement of your tongue, and how *roo* takes on a certain airiness! Now sound out these three syllables as a continuum: *oo-loo-roo*! It has a pattern, a rhythm – short, short, long, rather like an *anapaest* in poetry – as it must when you momentarily disrupt the *oo* stream with the consonants *l* and the *r*. But the consonants in this case, *l* and *r*, are 'soft' and, far from interrupting the flow, enhance it. The *l* and the *r* 'roll off the tongue', as it were, when coupled with the thrice-occurring *oo*, enabling the flow of *oolooroo* in a continuous stream without obstruction. You *would* interrupt the flow if you substituted, say, the hard consonant *k* (produced at the back of the throat rather than with the tongue) for *l* and *r*. To do so would involve much more effort, in that you have to block something off in the throat to make the sound *k*.

We need only practise listening to ourselves – which we never do – with a wonder-filled quietness and an expectation that there is something new to be found, and, in time, actually experience that the vowel *u* (oo) has a particular mood (surprise), different in kind from, say, *a* (ah) (wonder), which is originates further back in the body. Once these moods are *felt* it becomes a source of deep interest that this vowel *u* appears so frequently in the place names given by Australian indigenous people, in Wooloomooloo (NSW), for example, or Oondooroo (SA) and Ooloongathoo (QLD). Consonants are the plastic elements in speech, giving form to, materializing, things, whereas vowels give utterance to the inner life.

It was Barfield's particular genius to chart the evolution of consciousness through the history of the meanings of words (2000) and to demonstrate, contrary to conventional linguistic theory, that the further we go back in history the longer, not shorter, the words become (1967, 1984). We can gather from this that the distinction between speech and song was not as marked as it is now, so that the utterance of the time consisted of sung speech or spoken song. The words were *intoned*. One has the distinct sensation while listening to it that this music *in* speech is still there in the language of our indigenous people. Their song is contained within a very narrow pitch band; that is to say, there are no large leaps between notes, as one finds in the melody of Western culture. *Oolooroo* is eminently singable, the vowel making that possible.

The *spelling* of the word you have made is Uluru. The letters, *as* letters, give us nothing as aural experience. They are signs that freeze the temporal sounds in space. We have to practise a kind of selective deafness in order to form a word, by erasing the *experience* of the individual sounds; otherwise the progression from spelling to reading would be out of the question. No word is the letters comprising it. It is first an utterance, and, as utterance, its purpose is to make manifest. In making manifest, *modern* words detach things and events from their nexus in space and time, '… which puts the world, like a ball, in our hands' (Emerson 1907: 143). Naming things, insofar as we are non-participants in the naming, gives us power over them. They become things. Ayers Rock, as Uluru was once officially called, is conceived by many as a fossil, a dead object in equally lifeless space. It is a literalism, because we do not recognise 'that modern words that appear to us to have an exclusively material reference [rock] once also had an immaterial one'

(Barfield 1987: 42). Words used in this literal way are instruments of premature conclusion. They *contract* meaning. A rock is a rock is a rock.

At some point in our evolution, sound and meaning parted company. For pre-literate cultures, the *sound* of a word could not be detached from its meaning, or from the thing named. We can only guess, then, at the picture that rose up in the mind of the ancient Aborigine in the very uttering of the name, Uluru. We are speaking of a consciousness for which reality consisted of images, not things, a pre-conceptual consciousness 'that was figurative through and through; a kind of consciousness for which it was impossible to perceive *un*figuratively', and for which 'there is no such thing as a merely "outer" world' (Barfield 1987: 46). Uluru, for such a mind, is no physical object; it is the stopping place of a spiritual being, however implausible the *concept* spirit may seem to us now. Plausibility depends on accepting that language has adapted to earthly conditions, to the solid world.

I have undertaken this brief exercise in linguistic excavation in order to suggest that something of what I have described takes place for children everywhere as they are learning to speak. They are acutely attuned to the individual *sounds* of speech, or no intelligible speaking at all would be possible. They *listen*. It would be a mistake to attribute this act of attention to the ear. The ear, a physical organ, does not listen. A person listens; and it is a person who speaks. An understanding of this relationship between listening (giving heed) and speaking (giving voice) is vital to a full account of speech. Goethe (cited in von Lange 1992) evidently spoke of the ear as half a sense. The eye both sees and expresses. The ear hears, but expresses nothing. Its expressive counterpart is the larynx, to which it is connected physically by the Eustachian tube. Ear and larynx are thus articulated. This coupling may have little significance until it is realised that, contrary to popular opinion, the larynx is not moved by the air moving through or across it: it is self-moving. When listening to a speaker or a singer the larynx responds in sympathy, as it were, although we are not usually aware of it (except that one might occasionally notice a sore throat soon after listening to a forced voice at a concert, opera or lecture). The larynx is *intentional*. It imitates what the ear hears; but this imitation is no mere copying, the difference noted by Barfield when he wrote that we 'live in a camera [distinguished from harp] civilization' and, that 'it is already becoming self-evident to camera man that only camera words have any meaning' (1999: 54). If this imitative movement did not happen simultaneously with listening, children could never learn to speak. It might be said, then, that we hear with the ear but listen with the larynx, which, as living tissue, is capable of the most extraordinarily subtle and differentiated movements. The most evocative way I can put this is by way of musical analogy: the vocal chords play themselves by being played on. Speech is always accompanied by (usually unconscious) physical movement, or if not actual movement then movement *intentions*, which can be conceived as gestures. It is this movement-readiness that enables the larynx to 'sculpt' the formless air with its forms, which would otherwise remain gestures of the whole body. (The air, for its part, is always the bearer of something else, and surrenders itself for this purpose.) The larynx is a metamorphosed body, which has become itself by internalising, by becoming an inside, a voice. The phonemes are really creative gestures of the larynx, gestures that have first been *received*. 'For, strictly, it is language that speaks. Man first speaks when, and only when, he responds to language by listening to its appeal' (Heidegger 1975: 216).

We are in error if we think that speech, and by extension literacy, exists solely for the purpose of expression. It is what marks us as human. Every person who speaks, regardless of linguistic origin, not only gives form (to concepts and ideas) through speech, but is formed *by* speech. We speak and are simultaneously spoken. Heidegger wrote about our reluctance to consider this transformational power of the word: 'When this relation of dominance [language as the master] gets inverted, man hits upon strange manoeuvres. Language becomes the means of expression. As expression language can decay into a mere medium for the printed word' (Heidegger 1975: 215).

For children, speech and words are not a matter of routine. Since they have not yet become self-conscious observers, their listening is not a listening *to*, but a listening *with* the thing heard, an act of surrender. Listening *to* already presupposes a conceptual boundary between hearer and heard, an observer/observed confrontation. For children, that boundary is not yet a reality, and so they *metabolise* speech. They take its *qualities* into themselves unimpeded – for they cannot shut non-existent earlids – where it is transformed into ethical will. The moment we begin to speak about qualities we are in ethical territory. If the sonic world, the language and music that children digest, is as polluted as the physical environment, then we do their developing agency a grave injustice. I noted at the beginning that imagination cannot be inculcated or enforced. It can, however, be kindled from within. For we adults, this implies a certain reverence for the listener, the receiver, in the child, by exposing her to high order examples of speech and song as she is learning to speak and sing.

We usually ascribe image making to an inner eye; but there is an 'ear' of imagination as well, as great poets know. To a logician, the *sound* of a word is inconsequential; to a poet it makes all the difference, '"... poetically man dwells ..." says the poet' (Heidegger 1975: 216). It is to the poet in the child that we must attend, because the quality of the sound experience refines the power of observation.

Nobody of sound mind would deny the real need for literacy. I wanted only to show that there is more to literacy than meets the eye; there is also a literacy of aural experience, to which we turn a deaf ear at our human cost. We attach a word to a thing, and call it truth, while missing the truth that is the saying itself. Since we see ourselves only as users of words, we are far more interested in what we do with words than in what words can do. Selfless devotion to the word through the power of utterance to evoke images authorises us to reject the dead and deadening abstraction (language minus image, figure and motion) of managerialism that has spread like an invasive weed through government, corporate and educational sectors, and empowers us to create the conditions where every human meeting, whether conversation, lecture, seminar or concert, can become a ritual, a purposeful community of being. This is moral imagination in action. Reality principles are not worth much if they fail to ignite enthusiasm for a higher level of existence. We do not have to suffer the burden of normality. What if?

I leave the last word to Barfield in the Foreword to *Orpheus*: '… knowledge without imagination is not knowledge at all, but only a kind of cataloguing' and 'knowledge without love cannot be knowledge with imagination' (1983: 9).

References

Barfield, Owen (1967) *Speaker's Meaning*, Bristol, UK: Rudolf Steiner Press.

Barfield, Owen (1983) *Orpheus: A Poetic Drama*, West Stockbridge, MA: Lindisfarne Press.

Barfield, Owen (1984) *Poetic Diction: A Study in Meaning*, Hanover, NH: University Press of New England.

Barfield, Owen (1987) *History, Guilt and Habit*, Middletown, CT: Wesleyan University Press.

Barfield, Owen (1999) 'The harp and the camera', in G.B. Tennyson (ed.) *A Barfield Reader*, Edinburgh: Floris Books.

Barfield, Owen (2000) *History in English Words*, Great Barrington MA: Lindisfarne Books.

Camus, Albert (1991) *The Myth of Sisyphus and Other Essays*, Justin O'Brien (trans.) New York: Vintage International.

Donaldson, Ian. (2009) 'What have the humanities ever done for us?', in Helen Sykes (ed.) *Perspectives*, Sydney, NSW: Future Leaders.

Emerson, Ralph Waldo (1907) *The Works of Ralph Waldo Emerson*, A.C. Hearn (ed.), Edinburgh: W.P. Nimmo, Hay and Mitchell.

Heidegger, Martin (1975) *Poetry, Language, Thought*, Albert Hofstadter (trans.), New York: Harper & Row.

Polanyi, Michael (1969) *Personal Knowledge: Towards a Post Critical Philosophy*, London: Routledge and Kegan Paul.

Steiner, Rudolf (1992) *The Philosophy of Spiritual Activity: A Philosophy of Freedom*, Rita Stebbing (trans.), Bristol, UK: Rudolf Steiner Press.

Von Lange, Anny (1992) *Man, Music and Cosmos: A Goethean Study of Music*, Florence Hough (trans.), Sussex: Rudolf Steiner Press.

Winkler, Franz E. (1960) *Man: The Bridge Between Two Worlds*, New York: Gilbert Church.

PART 2

THE SOCIAL IN ECOLOGY

7

CREATIVITY COUNTRY

A journey through embodied space

Ainslie Yardley

Like light along a cable, sound down a wire or wind through the grass, new understandings ripple through us, flooding our entire body with a shiver of knowing that will change us forever. I can recognise this when it happens in another person. I can see it in their body, this passage of one way of being into another – this magical creative shift. When it happens to another they seem to be somewhere else. When it happens to me I seem to be somewhere else, other than where we usually are. Where do we go for this creative moment to occur? If we go somewhere, if creativity has both a moment and a territory, exists in time and space, in a country of a kind, where is this country between knowing and unknowing? The emerging place of all new thought, all new human construction – all new knowledge that is central to our existence as human beings. In many cultures, being in country describes having entered a place of belonging, of being purposefully within an environment to which one is spiritually connected and to which one has serious obligations, a collective space. It would seem to me that creativity is a space very like that – collective, serious, purposeful, of fundamental importance to us all.

The notion of creativity as country emerged for me, initially, as a metaphor to help conceptualise this phenomenon that seemed to be imbued with both spatial and temporal characteristics. To help explain how it was being in creative space myself, watching others go there, going there with them. It helped to conceptualise the 'rules' I felt, instinctively, applied. Creativity, like all countries it seemed, had borders, with 'laws' governing conduct within those borders, and conditions of exit and entry. Protocols. If certain procedures aren't followed we just don't get in. This notion of country became over time more than metaphor (Yardley 2000, 2004).

What kind of country brings a sparkle to the eye, a growth in stature, a delicious knowing shiver – a sense of having been backward and forward and in the middle all at once, in a place where the light, the sound, the wind of Self blows through – yours and everyone's. What kind of place is that? When you've been there it seems almost impossible to explain, and would be impossible to explain if we didn't have something to show of our travels there. Something to bring out that describes the nature of our journey. The yarns, the artefacts, the stories, the pictures and the maps of where we went, what we did there, what we learned – the material proof of our discoveries.

I have brought into my concept of creativity, with the greatest respect, a key element of what I understand (as a non-indigenous Australian) of the Aboriginal concept of being 'in country'. My understanding of being 'in country' (one's own) is of one being in or having returned to a place of belonging, an originating place, an environment in which one moves purposefully, to which one is spiritually connected and to which one has serious obligations – a collective space.

The phrase 'in country', is also used by the military, diplomats and officers of non-governmental organisations in a very different context, and with a very different tone. In this context, being 'in country' means having entered a place belonging to someone else, of being purposefully (that is, with the objective of doing a job of work) within an environment to which others are spiritually connected, and to which others have serious obligations, often very different from one's own. This is not the context in which I use the phrase in relation to creativity.

I make this connection with the Aboriginal concept of country not in any flippant or simplistic way, but to provide conceptual associations. Associations that can give greater depth of meaning to my representation of creativity as I consider it to be – a place of belonging, to be entered and returned to with seriousness and reverence. This conceptual connection reinforces the idea that there are ethical implications to 'being in country', both for those entering and those 'guarding' the borders. I imagine the border into creativity country as being quite a literal one, an actual boundary in time and space that we must cross with intent, in order to inhabit the imaginal world. We are, generally, within the constraints of everyday life, our own creativity country border guards, in possession of our own entry permits – except at those times when others (with the power to do so) take those freedoms from us and seek to limit the opportunities we have to inhabit the space and time of our imaginal world.

Becoming in the ripple – creativity and the body

When I was an undergraduate student (many years ago) I augmented my meagre income by working as a model for life-drawing classes. Two hours at a stretch of doing nothing other than striking a pose, keeping extremely still and being 'seen'. Of course, 'I' was not 'seen' per se, rather it was my body, as an 'object', that was seen or, more correctly, looked at intently to be interpreted – an unusual state of affairs, being so thoroughly seen and yet not seen. Depending on the length and configuration of the poses, it could be both a physical strain and a bore. To entertain myself, I would observe people as they drew, making the most of this wonderful opportunity to observe others' struggle with their creative vision. I would wager with myself which drawings would turn out to be the good ones, the dynamic, expressive ones. I rarely (if ever) lost a bet with myself. When they were good, I could perceive a change, almost like a ripple, pass through the person drawing, a physical 'sign' of some creative shift. This ripple of change seemed to visibly mark the creative moment, the emergence of something new into the world – something that could be seen and felt.

As observers outside an 'other's' body experiencing such a moment (someone drawing a powerful evocative image), we cannot know what the substance of the experience might be for them, but the change in them can be seen as it happens – an intake of breath, a sudden sparkle in the eye, a lengthening of the torso, a full, sweeping, energy filled gesture. Whatever this new awareness might be, it is accompanied by an alteration in the perceivable state of the body of the other, a growth in stature as the creative event occurs, a visible 'leaning toward' an idea to receive it.

When I model, I am the subject. I see you, but you are unaware of me seeing. You see my body as the object of your creative exploration, but you do not see 'Me' as a person. I watch you draw. I might see your body move with fluidity and grace, a light shining in your eye as your hand moves across the paper – if you are drawing well, if you are grasping something. I perceive the flow of energy from eye to body, to hand, to body. Your eye, my body – my body through yours and back again. I sense a continuous flow of energy between us until the essence of me (as body) is revealed, the nature of the exchange between us understood, the image captured. On the paper a likeness not of me, but of an exchange between two bodies about the nature of what 'body' is. I am confident the drawing will be 'good' because I have seen the ripple travel through you. The space that you and I have inhabited together as you drew and I was drawn seems to be not of this mundane world. We were for the duration both see-er and the seen, existing somewhere else.

I 'understand' something about you (and myself as witness) if I witness you transforming in a creative moment. I understand that you have experienced something deeply significant to you, without you saying anything. If creativity is manifest in the body, is lodged in the body, and this manifestation can be observed by me, and mean something to me about an 'other' – be seen in me by others – it must be, in some sense, relational, rather than the experience of an individual in isolation. This relationality stems from the 'phenomenological fact that we are always bodily in the world' (Van Manen 1990), sharing an interpersonal experience in the lived space and time of the world. When we enter creativity country we are bodily in that world sharing space and time, knowledge and ideas in the act of creation.

Creativity: Word and idea

Creativity as a word and an idea has become ubiquitous, devalued by faddish overuse and as a result less meaning-full in any context – popular culture, academia, the corporate world. Contemporary scholars have approached the phenomenon of creativity with some reticence and self-consciousness, reluctance even, over recent decades. Rarely touching on how we might define creativity, to whom it might belong or, most importantly, what the experience of it might be like.

This reluctance to consider the 'big picture' is intriguing in itself. Perhaps it is the product of increasing specialisation; perhaps it can be explained by the current obsession for pursuing field-centred certainties, in response to a widespread demand for certainties in all spheres of life. Perhaps it results from a discomfort with the indeterminate quality of the word itself, the difficulty of pinning down meaning[1] (Weiner 2000). It could hardly be claimed, however, that questions about the essential nature of creativity have been avoided because they have already been adequately answered. They clearly have not. This problem of beginning to talk about, let alone define, creativity is a perennial one.

In the last sixty years, research has tended toward a utilitarian psycho-social approach (Guilford 1954, 1959, 1967, Anderson 1959) that is focused on developing methods for measuring capacity; and later, in reaction to this approach, a shift towards identifying and studying systems-oriented models (Amabile 1983) and specific characteristics (or 'characters') of psycho-pathology (Arieti 1976) and genius (Getzels and Csikszentmihalyi 1964, Feldman et al. 1994), or creative 'innovation' and 'propulsion' models (De Bono 1977, Michalko 2001, Sternberg et al. 2002) – rather than toward an exploration of the more universal question of creativity's nature and purpose.

Csikszentmihalyi (Csikszentmihalyi 1996) considered the lived experience of 'creative flow', edging the discussion towards a more embodied approach, and Matthew Fox spoke from the theoretical wilderness, describing creativity as 'a place, a space, a gathering, a union, a *where* – wherein the Divine powers of creativity and the human power of imagination join forces' (Fox 2002). Questions about what creativity is or why the creative process repeatedly 'works' in a positively 'transformative way' have been largely ignored, other than by those writers with an interest in the place of creativity in the creation of meaning in disrupted lives, and the rewriting of self (Freeman 1993, Csordas 1994, Frank 1995, Becker 1998, Yardley 2000). The largely neglected phenomenological questions about creativity's nature and purpose cannot be usefully explored using psychometric testing (Guildford 1959), or by studying recognised exemplars of creative 'ability' (Feldman et al. 1994, Snyder, 2002). Studying the Stravinskys, Freuds and Einsteins of this world (interesting as the results of these studies might be) is to study simply that – exemplars of creative genius – not creativity itself. We need to step across the border into country and adopt an insider's view.

Creativity is commonly viewed as an attribute determined by the neurobiological matrix, a kind of genetic lotto win – a one in a million lucky break for the individual (Gardner, 1982, Csikszentmihalyi 1996). Creativity research (predominantly in the field of psychology) has sought to identify, the characteristics that make up the creative individual so that we can pick the talented innovator or the creative genius out from the crowd to ensure maximum advantage economically or strategically (Snyder, 2002).

The concept of 'gifted' individuals creates distinctions that are not particularly useful to us here – distinctions made according to assessment criteria I would like to set aside in this context in favour of a more expansive and inclusive premise. When travelling in 'creativity country' (Yardley 2004), everyone is gifted – when a child or an adult enters and journeys through the country of creativity, they do so with the intention of searching for and working towards the 'gift' of discovery, understanding and shared knowledge; and they are 'gifted' insights through the fruits of their labour (Yardley 2011). Entering creativity country entails work; seeks work; and seeks through that work to create artefacts that communicate insights, and map the journey undertaken. Creative products (works), new concepts and theories are determined (by those who make up their audience) to 'work' or 'not work' according to what they convey. Whether what we have learnt and crafted through shared journeys in creative space 'works' or doesn't work (for ourselves and for others as audience) depends on the depth and breadth of our explorations, and on how well the artefacts created communicate to others the insights that have been 'gifted' to us.

Individual creative attributes (talents) can be compared to 'talents' on the sports field – achievements to be appreciated and encouraged, but not confused with or placed above more universal experiences and outcomes such as the well-being gained from healthy physical activity and the enjoyment of collective physical achievement. 'Gifted' artists (and thinkers) may be the elite 'athletes' of the creativity field, capable of consistent high level achievement and insight, but creative acts are not exclusive to them. The patterning and mapping of our experiential and perceived world that we all engage in when we paint and draw, dance, create music, write poetry, stories, make plays, invent and experiment is a vital tool in understanding and shaping our social world and our place within it.

Creativity and consciousness

To begin to understand what creativity's 'moment' might consist of requires viewing from both internal and external vantage points, reflecting what is occurring within the individual body and in the environment external to it – as evidenced by the observable 'ripple' through the body as a creative act occurs. Considering creativity's relationship with consciousness is of primary significance in developing our 'insider's' view of what creativity might mean to every human being.

Core consciousness, neurobiologists tell us, is a faculty we share with other creatures. The faculty that provides 'a rite of passage into knowing' about our world and how to sustain ourselves within it – feed ourselves, keep ourselves warm and safe, recognise dangers, and nurturing relationships and environments (Damasio 2000). Extended consciousness (the faculty relevant to this discussion) builds on that capacity, permitting levels of knowing that can sustain human creativity and communication. The capacity that consciousness extends to creativity, and which creativity obligingly puts to good purpose, is the ability to gather stored memory and 'bring it into mind' in a narrative form. This capacity allows us to transform and combine images drawn from that 'wellspring of creativity', autobiographical memory (Damasio 2000: 24). Emotion, as Damasio explains, is the intelligence gathering mechanism, working away diligently in the background, unbeknownst to our conscious mind, recording all that we experience in the world, including the workings of our own interior, all the while laying down autobiographical memory. Damasio describes extended consciousness, the very core of human relationality and communication, as hinging on the ability to create an autobiographical record (that can then be brought into mind) in which one's sense of Self, the 'core you' is 'connected to the lived past and the anticipated future' (Damasio 2000). We make these connections by accessing autobiographical memory (brought to mind through feelings) and creating a continuously updated narrative in which we 'transform and combine images drawn from the repertoire of patterns and actions stored in memory' to discover new ways of doing things and to 'make new plans for future actions' (Damasio 2000, Yardley 2008). Extended consciousness grows across evolution and the lifetime of individual experience, collecting, adding to and reworking our sense of self in relation to the world. Creativity and consciousness are co-dependent. Without consciousness, creativity disappears. Without creativity, extended consciousness (knowledge of selfhood and our relationships to others and the world) would be an existential curse.

Ways of looking and seeing – interpretive and critical paradigms

On an everyday human level, 'bringing into mind' is also intimately associated with the 'word', as symbol-rich word-filled thoughts are translated into storyline and plot. These symbol-rich thoughts are filled with image as well, but the 'word' has become our primary method of ordering our mental contents, and transmitting and recording ideas, thoughts and feelings to others on a daily basis. Words exist for us to symbolise the concept or thing that they refer to in order that mental contents can be structured, communicated and remembered. 'Woman' symbolises a generic female human being, 'Nancy' symbolises, generally, a woman known by that name, or specifically my friend who lives in Samoa. 'Nancy-boy' on the other hand symbolises a concept – albeit a narrow prejudicial one. 'The word' ascribed to God in the Judaeo-Christian tradition is more than symbol. It is trans-substantive, transforming matter, becoming 'the thing' itself. God says, let there be 'Nancy' and there she is. If I were to say 'Nancy' with the full expectation that she would materialise, I would be considered naively misguided. I am expected to understand that the word is not the thing or the person that it symbolises. My relationship with the world is dependent on this understanding.

There are two quite normal circumstances, however, when that awareness is usefully suspended. When I dream – something I have no control over – anything can transform into anything. When I move through liminal space, where the temporary suspension of day-to-day awareness is safely undertaken (Turner 1969, 1974, 1982, Turnbull in Schechner and Appel, 1990)[2] and on into 'creativity country' to create – something I do by choice – anything can be transformed into anything. The suspension of one's day-to-day awareness of the difference between reality and the symbolic world is one of the keys to creativity. The 'word' has an uncommon function in the act of creation when our symbol-rich, word-filled thoughts are transformed into artefact or embodied in the performer. If I paint a portrait of Nancy, write her biography, choreograph the 'Nancy ballet', my symbol-filled thoughts about her (unknowable to others and obscured, in a sense, from myself) materialise, are brought into being, become something in their own right. My ideas about Nancy (the conflict in Iraq, genetically modified foods or the nature of the colour blue) are transformed and made real; thus something unformed and unknowable becomes known.

Creativity country is a complex and multidimensional space – a terrain much too challenging to be traversed in a single chapter, in a single book, if we are to come away with a strong sense of 'insider' knowledge. The intention in this contribution has been to consider the phenomenon of creativity country in the context of learning through place, nature, story and community, to draw attention to the significance of the embodied, relational nature and purpose of our journeys there, and to provide a discursive framework that recognises and supports the trans-disciplinary nature of the discussion that needs to take place.

Creativity is not simply an individual attribute – a lucky genetic lotto win. Creativity happens in time and space, in embodied space. A space we enter with our whole body/mind to undertake the creative work needed to maintain a state of dynamic equilibrium within our relational world. It is the space in which we think deeply, imagine new possibilities, make new discoveries, test new ideas and create new things. Everything that is generated in the human relational realm has its origins in creative space – and depends on our ability to go there.

Creativity and consciousness are intimately linked. If we take away consciousness, creativity vanishes. If we take away creativity and the ability to transform and shape our worldview, extended consciousness becomes a chaotic existential curse.

Creative space is the space of personal and cultural transformation, the space in which we 'lose ourselves' and abandon our inhibitions. In a relational sense it is the space from which sympathy, empathy and compassion emerge – the space we go to in order to imagine what other people's lives are like and walk for a time in their shoes. Caring and nurturing others requires creative space. Compassion and empathy are impossible without imagination.

There is an ethical imperative in relation to access to creativity country. No matter who we are or what our role in life might be, we do other people (and ourselves) an injury by preventing entry into creative space. We damage others and ourselves if we interrupt, ridicule or disrespect the need to listen, learn and explore; the need to reflect and contemplate; if we block access to useful space and time; if we impose rigid timetables and rules without reference to those whose lives will be governed by them; if we remove or restrict personal agency and creative control in people's lives; if we impose our will, our viewpoint, our moral code without being willing to enter creative space with people to share their life experience and understand their worldview.

The artefacts we bring out of creativity country (the things we create when we are there) include our autobiographical memory; our ancestral and cultural memory and identity; our aesthetic output (artworks, literature, drama); our scientific output and industrial design; our cultural, ethical and societal systems; our developmental programs and our individual self-development.

The more frequently we journey into creativity country and the longer we spend there, the higher the quality of our 'maps' and 'artefacts', and the deeper the shared understandings gained from them, will be. The complexity of each journey into country determines its dimension, the space and time it will require. Creativity requires 'duration' – the time and space it takes for ideas to 'incubate' and to be fully realised – the space and time between what is and what ought to be.

Notes

1 The English word 'creativity' (the noun) only came into use in the late nineteenth century, to fill the linguistic gap arising from emerging Enlightenment views about innovation in the arts and sciences – a presumption that the qualities that exemplified originality and invention were common to both these fields. A word was required which allowed a trans-disciplinary meaning to be expressed (Weiner 2000).

2 An idea first put forward by the anthropologist Van Gennep to describe the processes involved in ensuring effective rites of passage across diverse cultures (Van Gennep 1960). These processes depend on the establishment of a transition space (the liminal space) separate from all day-to-day social activities through which initiates move to complete their initiation.

References

Amabile, T. (1983) *The Social Psychology of Creativity,* New York: Springer-Verlag.

Anderson, H. (ed.) (1959) *Creativity and its Cultivation,* New York: Harper & Brothers.

Arieti, S. (1976) *Creativity: The Magic Synthesis,* New York: Basic Books.

Becker, G. (1998) *Disrupted Lives: How People Create Meaning in a Chaotic World*, Berkeley, CA: University of California Press.

Csikszentmihalyi, M. (1996) *Creativity: Flow and the Psychology of Discovery and Invention* (1st ed.), New York: HarperCollins Publishers.

Csordas, T. J. (1994) *Embodiment and Experience: The Existential Ground of Culture and Self*, Cambridge, MA: Cambridge University Press.

Damasio, A. R. (2000) *The Feeling of What Happens: Body and Emotion in the Making of Consciousness*, London: Vintage.

De Bono, E. (1977) *Lateral Thinking: A Textbook of Creativity*, Harmondsworth, London: Penguin.

Feldman, D. H., Csikszentmihalyi, M. and Gardner, H. (1994) *Changing the World: A Framework for the Study of Creativity*, Westport, CT: Praeger.

Frank, A. W. (1995) *The Wounded Storyteller: Body, Illness and Ethics*, Chicago, IL: University of Chicago Press.

Freeman, M. P. (1993) *Rewriting the Self: History, Memory, Narrative*, London and New York: Routledge.

Gardner, H. (1982) *Art, Mind, and Brain: A Cognitive Approach to Creativity*, New York: Basic Books.

Getzels, J. W. and Csikszentmihalyi, M. (1964) *Creative Thinking in Art Students: An Exploratory Study*, Chicago, IL: University of Chicago.

Guilford, J. P. (1954) *Psychometric Methods* (2d ed.), New York: McGraw-Hill.

Guilford, J. P. (1959) *Personality*, New York: McGraw-Hill.
Guilford, J. P. (1967) *The Nature of Human Intelligence*, New York: McGraw-Hill.
Michalko, M. (2001) *Cracking Creativity*, Berkeley, CA: Ten Speed Press.
Schechner, R. and Appel, W. (eds) (1990) *By Means of Performance: Intercultural Studies of Theatre and Ritual,* Cambridge, MA: University of Cambridge Press.
Snyder, A. (2002) *What Makes a Champion!: Fifty Extraordinary Individuals Share Their Insights*, Camberwell, VIC: Penguin Books.
Sternberg, R., Kaufman, J. and Pretz, J. (2002) *The Creativity Conundrum: A Propulsion Model of Kinds of Creative Contributions*, New York: Psychology Press.
Turner, V. W. (1969) *The Ritual Process: Structure and Anti-Structure*, London: Routledge and Kegan Paul.
Turner, V. W. (1974) *Dramas, Fields, and Metaphors; Symbolic Action in Human Society*, Ithaca, NY: Cornell University Press.
Turner, V. W. (1982) *From Ritual to Theatre: The Human Seriousness of Play*, New York: Performing Arts Journal Publications.
Van Gennep, A. (1960) *The Rites of Passage*, London: Routledge & Kegan Paul.
Van Manen, M. (1990) *Researching Lived Experience: Human Science for an Action Sensitive Pedagogy*, Albany, NY: State University of New York Press.
Weiner, R. (2000) *Creativity and Beyond: Cultures, Values, and Change*, Albany, NY: State University of New York Press.
Yardley, A. (2000) 'Creativity country', paper presented in symposium at University of Western Sydney, Hawkesbury Campus, January.
Yardley, A. (2004) 'Creativity country: A study of the phenomenon of creativity in relation to disrupted life', unpublished thesis, University of Western Sydney.
Yardley, A. (2008) 'Living stories: The role of the researcher in the narration of life, *Forum Qualitative Sozialforschung* (*Forum: Qualitative Social Research*), 9(3) Art. 3, http://nbn-resolving.de/urn:nbn:de:0114-fqs080337, accessed 4 February 2011.
Yardley, A. (2011) 'Children as experts in their own lives: Reflections on the principles of creative collaboration', *Child Indicators Research Special Issue*, New York: Springer Publishing.

Further reading

Berndt, R. M. and Berndt, C. H. (1974) *The First Australians* (3rd ed.), Sydney, NSW: Ure Smith.
Bruner, J. S. (1986) *Actual Minds, Possible Worlds*, Cambridge, MA: Harvard University Press.
Cowan, J. (1994) *Myths of the Dreaming: Interpreting Aboriginal Legends*, Bridport, UK: Prism Press.
Dennett, D. C. (1991) *Consciousness Explained* (1st ed.), Boston, MA: Little, Brown & Co.
Dennett, D. C. (1996) *Kinds of Minds: Towards an Understanding of Consciousness*, London: Weidenfeld & Nicolson.
Dennett, D. C. (1998) *The Creation of Creativity*, www.centreforthemind.com/publications/newsletter/Mindspace1-1p4.cfm, accessed 17 May 2002.
Dissanayake, E. (1988) *What is Art For?* Seattle, WA: University of Washington Press.
Eliade, M. (1973) *Australian Religions: An Introduction*, Ithaca, NY: Cornell University Press.

Fox, M. (2002) *Creativity: Where the Divine and the Human Meet*, New York: Tarcher.

Gardner, H. (1985) *The Mind's New Science: A History of the Cognitive Revolution*, New York: Basic Books.

Milgram, R. M. (1991) *Counseling Gifted and Talented Children: A Guide for Teachers, Counselors, and Parent*, Norwood, NJ: Ablex Publishing Corporation.

Panel on Human Factors in the Design of Tactical Display Systems for the Individual Soldier, N. R. C. (1997) *Tactical Display for Soldiers: Human Factors Considerations*, http://books.nap.edu/books/0309056381/html/7.html, accessed 12 May 2004.

Runco, M. A. (2007) *Creativity Theories and Themes: Research Development, and Practice,* Burlington, MA: Elsevier.

Stokes, P. (2006) *Creativity From Contraints – The Psychology of Breakthrough*, New York: Springer.

Yardley, A. (2008) 'Piecing together – A methodological bricolage', *Forum Qualitative Sozialforschung* (*Forum: Qualitative Social Research*), 9(2) Art. 31, http://nbn-resolving.de/urn:nbn:de:0114-fqs0802315, accessed 9 February 2011.

Yardley, A. and Bailey, J. (2007) 'Visiting creativity country: A policy-makers travel guide', *Applied Theatre Researcher/IDEA Journal*, 8, Brisbane, QLD: Griffith University Press.

8

TURNING ON A SIXPENCE

Creating a learning and systems thinking foundation for community participation

Sally MacKinnon

A story emerges

Recently, I realised that I am a gardener. After filling two decades of my adult working life with education, public relations, community organising and copywriting – generally focused on environmental sustainability – it's a surprising realization to recognise that in fact I have grown into a gardener these past two years. Not just any sort of gardener, mind you, because while I enjoy tending my household's own backyard permaculture garden, I'm not actually much of a literal gardener.

I am a community gardener.

I have realised that I perceive communities as gardens – as living, evolving, self-organising organisms. Living entities that grow, seed, produce, entangle, disentangle, wither, collapse and regenerate in continuing, ongoing ways. I have realised that communities, especially my own home community, don't depend on my professional expertise as an educator, trainer, communicator and organiser to make them work, though I can sometimes contribute to their humming, buzzing life if I listen carefully, look closely, learn deeply, care greatly and recognise myself as a part of this living system (not separate or superior to it).

Over the past two years I've been rebuilding my life from the soil up. Very quietly and gently with lots of time and space to sit, reflect, observe, listen, hear and see. I've let the old driver go and the new gardener grow. And so this little story about learning and community arises from the very heart of my own experience.

Bugged

During this time of reflection I've been bugged by a sense of frustration about community and communication, particularly in the areas of community consultation, community education and community engagement. In 2004 I completed a PhD thesis that was highly critical of community consultation because of its inability to meaningfully engage, innovate and enable real participation and real change. In that research I discussed the potential of grassroots, community-driven, multiparty negotiation to create new ways of making social and structural change from the ground up and from a base of learning in and from action

(MacKinnon 2009). The centrepiece of this research was the story of the 1996 Cape York Peninsula Landuse Heads of Agreement, a remarkable agreement brokered between the graziers, greens and Aboriginal people of the Cape. The agreement exemplified how diverse stakeholders (some who were enemies) could work together to protect and restore the quality of the region's natural environment, communities, cultures and economy.

In 2008 I was an active participant in a local government whole-of-community visioning project – a well publicised, much-awarded, thirty-year visioning process – as a member of the project's community advisory committee. While I'm not sure what that council thought I could bring to the table that would be of use to the project, what emerged for me during that year was a niggling, annoying, persistent bug that refused to be content with a community consultation approach based on the traditional marketing and communications machine – one that pours glossy information into the 'empty vessel' of community and asks simplistic questions of generally busy, time-poor people.

I must make clear that this is not an attack on the very skilful and capable council marketing and communication team, nor indeed on the council itself. This is a much broader bugbear about the traditional marketing and public relations approach that underpins most community consultation in Australia. It is an approach that, for some years, appears to have been reaching its limits of effectiveness, with evident levels of fatigue, disengagement and even mistrust setting in throughout communities whenever they're consulted by government or big business, the two most common proponents of community consultation.

During 2008, the community advisory committee reviewed substantial amounts of community feedback from public forums and questionnaires and waded through many expert presentations and papers in order to help draft the community's thirty-year vision, themes, principles and targets. Interestingly, right at the end of 2008 when the committee was due to go back to the community with its ideas, the whole committee began to feel stuck – at an instinctive, gut level we committee members all seemed to recognise that the glossy marketing brochures and website weren't in and of themselves going to tell an engaging, compelling and open-ended story that would catalyse the good people of the city to co-create their own future.

It seemed the whole committee was bitten by that pesky bug, but as a group, we didn't quite know what to do about it. So I suggested we garden, that we look around at our world in a different way – through a learning and storying lens.

Interesting times

There is a lot going on in the world these days. Global economic meltdown (to use the media's own language) reflected in extraordinary financial market freefalling around the world and on-the-run federal government cash management. Global warming and climate destabilisation made visible in phenomenon such as glacial meltdowns at the poles and chronic water scarcity throughout much of Australia.

Roller coaster petroleum prices and talk of oil shocks and peaks that affects food prices in affluent nations and food availability in the most vulnerable countries of the world.

Those on the margins, the ones dubbed the 'loony left' and 'tree huggers' by media stereotyping, have been banging on about such global possibilities for a while now. But in the past few years, highly credible political scientists like Professor Thomas Homer-Dixon (2007), renowned climate and natural resource scientists like Dr Graeme Pearman and Dr John Williams, and even corporate leaders like Ian Dunlop (2006) have joined the chorus, expressing dismay not only at the speed of these seismic ecological, economic and social shifts but also with their convergence. Our governments, businesses, communities and households are not just faced with one big issue at a time, but with a cocktail that mixes everything together in a bitter and highly synergised pill.

What are we to do when we bring these interesting times into perspective at a regional, city, community and local level? Albert Einstein was onto things all those years ago when he said that the level of thinking that created significant problems was *not* the level of thinking required to solve them. He was referring to the fact that our world is a very large, complex system made up of many interlocking, connected, smaller systems. He was suggesting that if we want to do more than continually create more problems out of one-track, linear, mechanistic thinking and 'solutions', then we need to bring many minds, many ideas, much creativity and great depth together in the way we think about and act in the world.

When I think about the complex and interconnected challenges that are facing humanity, I usually run in fright to my backyard garden and get my hands into that volcanic Beechmont soil. I look at the profusion of seedlings, flowers, fruiting and seeding vegies – greens, beets, herbs, beans, potatoes and so on – and I hear an amazing hum and buzz of pollinating bees and insects. I comprehend an extraordinary layering of relationships, connections, knowing, functioning, producing, composting, exchange, reciprocity, synergy and growth going on without a single leader, book, brochure or word spoken. Complexity and knowledge and activity and solutions are all around in my backyard garden. Knowledge and innovation and creativity are distributed everywhere – throughout nature and throughout human communities.

If we imagine for just a moment that our communities are gardens – living social systems – and that we are gardeners, then perhaps we will discover that we already have the means to think about and engage more meaningfully with our interesting times, if we simply change our perspective.

Paul Hawken (2007), ecologically sustainable business commentator and author, discusses the notion of distributed and localised knowledge and leadership in his book *Blessed Unrest* when he says, 'Since one person's knowledge can only represent a fragment of the totality of what is known, wisdom can be achieved when people combine what they have learned …' (21).

Organisational learning researcher and author, Peter Senge (in Webber 1999) suggests that we need to take off our mechanic/driver/controller glasses that assume the world is linear, simple, controllable, 'fixable', and that the answers lie in the hands of just a few strong leaders. Senge suggests that organisations and communities *are* like gardens – complex, self-organising, organic and evolving organisms that contain many leaders, much knowledge, multiple connections and relationships, and many layers of innovation and creativity. He suggests that our most important task *is* to become gardeners who can find and support and nurture these leaders, knowledge, connections, relationships, innovation and creativity.

If we become gardeners I think we will also naturally evolve into learners and begin to regard our communities, organisations, businesses and households as evolving sites of learning, vitality and innovation. We will no longer see them as 'empty vessels' that need 'the right' information poured into them, or as machines that need fixing. We have an opportunity to change our lens on the world from one of mechanical control to one of co-learning in and from complexity and action.

Reaching the limits of marketing and communications

I would like to suggest that the methodologies of public relations, marketing and media communications have been the assumed foundation for community consultation and community engagement for too long. According to Australian environmental researcher Sharon Beder (1997):

> modern public relations dates back to at least the 1930s, when Edward Bernays 'convinced corporate America that changing the public's opinion – using PR techniques – about troublesome social movements and labour unions, was far more effective than hiring goons to club people'. Bernays had worked for the wartime propaganda commission in the US, and wrote up his ideas in articles with titles like *Manipulating Public Opinion* and *The Engineering of Consent*, which described the 'application of scientific principles and tried practices in the task of getting people to support ideas and programs'.
>
> (107)

Even within the environment and sustainability movements that I've been part of since 1989, public relations, marketing and media communications have been a central base upon which the movement has built much of its public education and campaigning. It is a paradigm that assumes 'they' (the public, the politicians, the business leaders etc.) do not know or understand the big issues, and if we can just feed them the right information or knowledge in an appealing and persuasive way then they will hear, understand and change – it's the old 'empty vessel' model of education and public communication again.

I would like to suggest these models are now singularly ineffective because:

- 'They' are not empty vessels – people are complex and evolving beings who embody a great deal of knowledge and are constantly engaged in learning processes in conscious, unconscious, formal and informal ways within their lives.

- 'They' are people who have diverse values, beliefs and roles and come from a rich diversity of backgrounds and cultures that greatly colour the way they see and engage with the world.
- The big issues and the limits to growth that human civilisation is now facing are complex and interconnected and cannot be addressed in linear, simplistic, single message ways.
- Many people are now skeptical about and literate in the ways of marketing and public communications and simply do not trust or, indeed, engage with these processes and messages any more.
- Most people are extremely busy in their lives of work, family, community, education and recreation and don't have or make the time to engage with marketing and PR (which they're already mistrustful of).
- Most people have little trust in government, big business, lobby groups and activists – or their marketing messages – because they sense there are hidden agendas and a lack of authenticity and credibility.
- People have access to an increasing diversity of information, well beyond traditional mainstream mediums, and increasing numbers of them seek out alternative sources of information as their needs arise.

The need for new stories

In these interesting times we are all in need of new stories that help us more clearly recognise our challenges and opportunities in ways that lead to cohesive, joined-up thinking and action. We need stories that are inspiring, energising, optimistic, compelling and authentic. We need stories that not only help us to reflect and learn but that also inspire us to create together in a process of innovation, engagement, experimentation and growth. Paul Hawken (2007) says that humankind *can* create new narratives and that:

> a society capable of naming itself lives within its stories, inhabiting and furnishing them. We ride stories like rafts, or lay them out on the table like maps … because stories are greater than we are, their capacious narratives give us wiggle room to dream … our families and communities connect us to the old and new stories, and guide us to lean into the light.
>
> (25–6)

Let's remember that stories are created collaboratively between the writer and the reader, and that stories can be told in many ways and through many vehicles: music, images, text, art, dance, online, in videos, through living examples, in forums, celebrations, meetings, around kitchen tables, in workplaces, in classrooms, playgrounds, in the bush, on the beach and in the old-fashioned book, just to name a few.

Good learning foundations, processes and stories can bring people together in ways that are unifying but not stifling. They don't demand agreement of opinion (surely an impossible task), but they do help to create alignment around an evolving, co-created vision or intention. They also allow the creation of a diversity of activities and pathways towards that aligned vision.

Good stories, when created and developed collectively, also have the potential to 'solve for pattern', Wendell Berry's (1981) marvellous term that refers to solutions that address multiple, connected, systemic problems (without creating further problems), instead of just dealing linearly with one symptom at a time. Berry, a farmer, poet and systems thinker suggests, for example, that sustainable food and farming has the potential to address many serious issues simultaneously, including: soil and water health, carbon storage and greenhouse gas reduction, worker health and safety, productivity and yield, water conservation, biodiversity, food nutrition and local economic prosperity. For Berry, sustainable agriculture is more than a linear 'fix', it is an evolving, systems-based approach to some of the biggest issues of our time. It also requires more than top-down leadership to implement. It is a place-based, localised, creative and evolving framework, philosophy and act of creativity arising from the grassroots.

The role of the community gardener

One of my favourite words in the English language is 'desireline'. I understand it comes from the landscaping fraternity and describes the informal pathways pedestrians make when they step off concrete footpaths and walk across lawns and grass. They are the routes that people take or make based on their needs and desires to get somewhere via a shorter or perhaps more scenic path and they tend to become visible over time as more feet walk them across the ground.

If we continue with our 'community as a garden' thread, I believe there also exist webs of invisible desirelines within our communities that connect ideas, knowledge, needs, aspirations and leadership with people, projects, organisations and businesses. If our communities are already living systems and sites of learning and innovation, then one of the most important jobs of local government, in particular, is to find those community desirelines and support their ongoing evolution, because these represent what our communities value at their core. I also believe it is important and possible for community desirelines to grow and function alongside the more formal structures of governance, though if we are to be community gardeners, we must be careful not to put up 'keep off the lawn' signs, or concrete over desirelines with so many layers of bureaucracy and structure that they become too rigid and uncomfortable to be useful to community any more.

During the 1990s, economic development researcher and practitioner Ernesto Sirolli (1999) developed a framework that aligns well with our metaphor of the community as a garden. Called 'enterprise facilitation', Sirolli's framework focuses on the development of local economies and small to medium enterprises that are fuelled by peoples' own passions and aspirations. Economic development officers are reinvented as enterprise facilitators – community-based networkers whose job involves doing nothing until people come to them with their business aspirations. Then their role is to listen deeply to understand those aspirations and find the connections, information, expertise and leadership within the community to enable those aspirations to evolve into thriving, meaningful business enterprises.

In suggesting a learning lens for community engagement it's useful to turn Sirolli's enterprise facilitators into learning facilitators: people who are great listeners, observers and connectors. People who have the ability to go into communities to find the local leaders, innovators, desirelines and needs and who can help connect them up. Facilitators who have the skills to help people and organisations reflect upon and tell their stories, then amplify those stories across communities so others can be inspired and activated and the learnings made contagious. And they also need to have the capacity to connect the grassroots with more formal governance structures so that real partnerships are born between government and community.

Supporting new growth

Something important needs to change if the government and community visioning and action projects are to become meaningful for people and communities. Our communities need and deserve more than a one-size-fits-all, top-down vision. Our communities are already well-stocked with effective local leaders, teachers and visionaries. They are already home to many effective, inspiring, locally relevant innovations too, from outstanding neighbourhood building and family support programs that are pulling down fences and raising roller doors (Northern Gold Coast Communities for Children Program 2008), to internationally recognised ecologically sustainable businesses. These programs, businesses and people arise from the ground up to meet the needs and aspirations of local areas in meaningful, authentic and appropriate ways.

If we want to support whole-of-community visioning and transitioning, we can do so in low risk, cost-effective ways simply by collaborating and partnering with the innovators already living a sustainability vision. We can learn from them and help make their stories accessible and contagious. We can map them and celebrate them and enable their work to grow and connect. In doing that we help community visions grow and expand in ways that work, and work from the inside out rather than being imposed from above. Naturally, there is also room for new start-up signature projects too, that can be tested initially at small, localised scales and grown or adapted as we learn from them.

In the end, I don't believe that community engagement built upon a lens of learning is a radical shift at all. Like the propensity of life to turn on a sixpence and our inbuilt human capacity for relationship, it can happen in an instant if we shift our perspective to the local level, to the learning level, to the recognition that our communities are living systems – gardens if you will.

References

Beder, S. (1997) *Global Spin: The Corporate assault on Environmentalism*, Melbourne, VIC: Scribe Publications.

Berry, W. (1981) *The Gift of Good Land*, Berkeley, CA: Counterpoint Press.

Dunlop, I. (2006) 'Unholy trinity set to drag us into the abyss', *Sydney Morning Herald*, 16 October 2006.

Einstein, A. http://www.quotationspage.com/quote/23588.html, accessed 7 February 2011.

Hawken, P. (2007) *Blessed Unrest: How the Largest Movement in the World Came into Being and Why no one Saw it Coming*, New York: Penguin Group.

Homer-Dixon, T. (2007) *The Upside of Down: Catastrophe, Creativity and the Renewal of Civilization*, Melbourne, VIC: Text Publishing.

MacKinnon, S. (2009) *Expanding Green Strategies: Creating Change Through Negotiation, Germany:* LAP Lambert Academic Publishing.

Northern Gold Coast Communities for Children Program (2008) 'Behind the Roller Door', DVD.

Sirolli, E. (1999) *Ripples from the Zambezi*, Gabriola Island, BC: New Society Publishers.

Webber, A. (1999) *Learning for a Change*, Fast Company, http://www.fastcompany.com/magazine/24/senge.html, accessed 9 February 2011.

9

INTEGRATING SUSTAINABILITY
Towards personal and cultural change

Jasmin Ball and Kathryn McCabe

Introduction

Humans have the innate capacity to live sustainably. However, within the current industrial culture this capacity is being disabled. This chapter illuminates some problematic implications of the industrial world's responses to sustainability by highlighting the broader impacts of these responses personally, socially and environmentally and, in doing so, proposes some of the psychosocial root causes of unsustainability inherent in current ways of living. Lastly, this chapter uncovers pathways for developing and integrating sustaining ways of living by exploring conditions that enable critical self-reflection, supportive relationships and ongoing personal healing.

Sustainability today: A critical reflection

Industrial culture continues to imply that we are living or striving for the 'good life' (Dittmar 2008: 1–15). We are often told that industrialised countries have some of the highest standards of living in the world. Yet so many of us are somewhat forced into the reality of working hard to survive, having little free time, being in debt and feeling exhausted. This current model of living is not sustainable. It's built on progress, growth, competition and individuals buying into the cultural representation of the 'good life'. The focus on the cultural ideal of the 'good life' draws attention away from all the symptoms that clearly show that our way of life is not working for most people or the ecosystems that support us. For example, why are so many people depressed (Beyond Blue 2010)? Why are more than one million people committing suicide globally every year (World Health Organization 2010)? Why are cancer, obesity and other serious health issues increasing exponentially (WHO 2010)? Why are some of us consuming goods that we don't need (Dittmar 2008)? Why do many feel a lack of meaning or purpose in their lives (Fien et al. 2009)? Why are ecosystems being destroyed everyday (International Union for Conservation of Nature 2010)? Is there a better notion for a 'good life' that we can create individually and collectively? Sustainability can be seen as an opportunity to engage with personal and cultural change to create a form of a 'good life' that truly enriches you, others and life on this planet.

The significance of sustainability has emerged in response to industrial society's devastating impact on natural resources that have led to implications for maintaining our excessive lifestyles. The newfound pertinence of sustainability has led it into the media, political agendas, educational curricula, industry protocols, consumer products and the general public's consciousness. As a result, interpretations, representations and associated practices are even more diverse. The issue, however, is that industrial culture has simplified and commodified sustainability. It primarily presents sustainability as involving a mixture of purchasing 'greener' alternatives, preserving and rehabilitating the environment, resource and waste management. As explored throughout this chapter, these approaches are limited and often fail to address the underlying causes of unsustainability.

Understandably, there are mixed feelings amongst the public about sustainability. There are plenty of people that perceive 'going green' as a threat, a conspiracy theory, 'touchy-feely stuff', an expensive lifestyle choice, too hard, giving up their way of life and a return to primitive-like living. It's not surprising that many people can feel bored, sceptical, guilty, overwhelmed, and helpless or 'switch off' when exposed to matters relating to sustainability. These are valid responses to the cultural representations of sustainability, and indeed to the reality of the crises facing life on Earth.

These individual responses to sustainability have also contributed to the development of the cultural approaches (outlined in Table 9.1 below). Cultural approaches also influence individual responses – there is a continuous interactive relationship between individual responses and cultural approaches to sustainability. Some of the current approaches and responses that industrial culture has adopted to address sustainability are listed in Table 9.1. Many may find themselves applying some of these approaches – that does not make us bad or stupid people! Most of these approaches/responses in Table 9.1 are well intentioned, often unconscious and some simply need to be readdressed in terms of desired outcomes. In highlighting these approaches and responses, the intent is to identify their implications on a personal and social scale. Furthermore, by drawing attention to the shortcomings of these approaches, it is hoped that we can address some of the underlying causes that relate to sustainability, understand complex interrelationships and as a result enhance our capacity to create deeply sustaining ways of being.

Rather than focusing on each example in Table 9.1, a variety of themes can be extrapolated from the outlined approaches. Whilst some can be seen to play a role in working towards particular aspects of sustainability, they often create further problems.

What is common amongst the approaches featured in Table 9.1 is that they do not give space for our personal experiences and wisdom. Often the people directly experiencing an issue have a wealth of knowledge and a depth of understanding crucial for addressing it (Peavey 1994: 83–90). Yet many people in our culture are depending on our governments, the 'experts' and the latest technologies to solve issues concerning them. By handing over our power to those in leadership positions we become susceptible to their imposed agendas, often to the detriment of our own (Komives et al. 1998: 8–82). Our own wisdom and experience is often exactly what is required to create contextually appropriate responses that are more likely to work effectively with the complexity of the issue.

As a facilitator I (Kathryn McCabe) have found using processes that create an opportunity for individuals to harness their own wisdom generates the most profound shifts in their understanding, motivation and action. Whether my work is with communities coping with ecological collapse along river systems in Australia, corporate teams seeking to embed sustainability into their business practices, young people wanting to make a difference or adults wishing to ignite their passion for life, these enabling processes are often transformational. Allowing their personal agenda to emerge can be very liberating and empowering; an experience that then tends to flow into other areas of life, especially with ongoing support.

Many of the current 'solutions' for sustainability are often seen as end points; either as a single answer to solve a problem or more often as sitting within an equation of answers, e.g. 'Dams + Recycling + Desalination + Water efficiency = Water 4 Life' (Sydney Water 2010). This approach limits the focus to a few areas, usually on the environment as a subset of the economy. Furthermore, this 'solution' fails to address the impacts of these actions, for example, that building dams disrupts ecosystems and interrupts the naturally evolved flow of water through the land, water recycling uses harmful chemicals, desalination uses immense amounts of energy and waste by-products include toxic brine. Creating a border around 'the environment' fails to perceive the interconnectedness of all 'systems' and therefore the affect they have on each other. The result is often a 'win' in one area of focus, albeit temporarily, whilst other interconnected aspects are worsened as a result. Hill (2009a: 6) comments that, 'Common interventions that use magic bullet, instant high-powered solutions, delivered by conspicuous heroes … usually eventually have numerous side-effects, few in the population are beneficiaries, and any positive outcomes are generally temporary, i.e. not sustainable.'

Thus, it is essential that the 'solutions' devised to foster sustainability are understood as sitting in a broader interconnected web of relationships. Through becoming aware of the interrelationships between things, our perspective of the world can change, and we increasingly begin to realise that in 'open' systems, everything is understood to influence and in turn be influenced by everything else (Thurling, personal communication, 5 March 2010). This is one of the most fundamental underlying features of the social ecology perspective (Russell 1991: 126–9, Woog and Dimitrov 1994: 1–5, Goff 1997: 39–42, Hill 1999: 199–201, Camden-Pratt 2008: 6–7), and plays a pertinent role in integrating sustainability into all areas of life. In this light, small, localised actions can have far-reaching impact.

Another theme that can be drawn from Table 9.1 is the emphasis placed on working with issues at the symptomatic level. Working at a purely symptomatic level with any issue keeps responses trapped in an unending cycle of problem–solution. Treating only the symptom leaves the underlying malaise to generate

Table 9.1 Dominant approaches and responses to sustainability

Approach/response	*Explanation and examples*	*Problematic implications*
Substitution	Creation of 'green' products and services that largely serve to substitute for the impact of our current lifestyle, e.g. buying a better mileage car but continuing driving, solar panels, eco-products.	Maintains the status quo. Avoids reflection on unnecessary consumption. Resource use not addressed. Justification for business (and pleasure) as usual.
Compensation	Systems and tactics that seek to compensate for unsustainable action, e.g. donating to charity, carbon offsetting and emissions trading scheme.	Creates a facade of sustainability. Serves to justify shallow, feel-good, 'I've done my bit' action.
Informational/ educational	Giving the public information and/or prescriptive guidelines for being sustainable based on the perspective that all that is needed for change is information. For example, if people know that littering, driving, flying is harmful to the environment they won't do it. In '10 simple steps you can change the world'.	Often delivers short-term, de-contextualised change. Disempowered reactive responses are strengthened. Cognitive dominance reinforced, emotional responses devalued. Conflicting information often leads to inaction, scepticism, mistrust, being overwhelmed.
Manipulation	Manipulative tactics used to create a desired outcome, e.g. fear mongering, imposed political agendas, financial incentives, auditing, green washing, green marketing, sensationalism, political leaders instilling a false sense of effective action.	Often leads to being overwhelmed, inaction, disempowerment, mistrust, compliance. When we are manipulated we often manipulate others, leading cyclically to further personal and social ills. For example, depression, trauma, violence and compensatory behaviours (Hill, 2009a: 6–7).
Problem solving	Waiting until a problem emerges and then devising solutions that do not address root causes. Reactive responses to undesirable situations. For example, conducting research, fighting against climate change, fighting for peace, protesting, foreign aid and taking a painkiller for a headache.	Limits creativity, vision and autonomy. Often short-term 'solutions' and mostly fails to deal with the deeper interconnected causes of the problems.
Marginal/ fragmented	A narrow perspective and approach to a sustainability issue. For example, sustainability only in terms of the environment, only about behaviour management, only about educating people.	Mostly results in restrictive and narrowly focused outcomes. Mostly fails to engage with the 'big picture' impact.

Approach/response	*Explanation and examples*	*Problematic implications*
Deferring power	Relying on ‘higher’ powers to take charge of the situation. For example, government/scientists/God/Gaia knows best. Technology will save us. Instilling power in the ‘experts’.	Delays action. Disempowering. Strengthens status quo.
Denial	Personal denial and media/government/corporate manufactured doubt. For example, ‘She’ll be right mate’, ‘It’s a conspiracy’, ‘It’s not happening’, ‘The government won’t let it happen’, ‘They’re still researching’, ‘There’s not sufficient proof’.	Delays and/or avoids personally meaningful responses.
Projection	Projecting the issue onto others. For example, I can’t/won’t do anything until the government does. The rich or the poor are the problem.	Incapacity to accept personal responsibility for our actions.
Justification	The endless excuses/justifications we can make to continue our lifestyle. For example, ‘I need/have to drive my car to get to work’, ‘I need a new iPod because my current one is broken’, ‘I’m too busy right now’, ‘It’s all too hard’, ‘I don’t know what I can do’.	Personal, organisational and cultural transition to sustainable practices delayed or avoided.

new manifestations. This approach is evident throughout a number of approaches critiqued in Table 9.1, such as substitution, compensation, informational/educational and problem solving. For example, much emphasis is placed on waste management by way of reducing, reusing and recycling, which of course is part of the solution. However, little emphasis is placed on addressing the deeper, often emotional, causes of unnecessary consumption. As a consequence, a majority of 'sustainable' solutions have a Band-Aid effect, failing to tackle the actual root causes. Maiteny (2009: 178) cleverly depicts these points:

> What we prefer to see as the 'problems' are actually ways of unconsciously avoiding these deeper causes within ourselves. These are the real problem, but this is just too excruciating to admit. Unless we do so, however, we will continue, ever more urgently and uselessly, to rearrange deckchairs on the ecological Titanic as it sinks deeper into the ocean.

At the same time, Band-Aid approaches often generate a facade of sustainability producing a misleading sense of success (Hockachka 2005: 12–14), meanwhile, the planetary crises continue.

While seeking greener substitutes can function to reduce the impact on the planet, it's essential to understand the broader implications of substitute choices. Indeed, substitution is currently one of the most common sustainable initiatives being encouraged, while little focus is placed on how resources are used or on general consumption behaviour. It would appear that switching to solar power can limit carbon emissions and reduce damage from mining fossil fuels for coal-produced energy. Yet, to use an oversimplified example, if solar power is used to cut down forests then little has been done to live sustainably from an integrated perspective. Thus, it is not the product itself, it is how it is used that creates the outcome. Why we consume and how we consume has deeper underlying causes that are often emotionally related. By purchasing 'green' substitutes, we often sidestep deeply questioning unnecessary consumption.

In a culture that values consumption, progress and assimilation it is probable that initiatives developed to tackle sustainability support these values. As 'modern' culture identifies strongly with a materialistic, resource-heavy way of life, it is understandable that individuals would avoid questioning their lifestyle choices and underpinning worldviews. Consequently, current 'sustainable' initiatives, such as substituting one product for a 'greener' alternative, are confined within the current paradigm of cultural norms. It's essential that we more deeply question the functions of industrial culture, otherwise, most 'sustainable' initiatives will fail in addressing the underlying root causes.

Therefore, all approaches outlined in Table 9.1 are predictable and appropriate responses within the context of the cultural norms of industrialised society. These cultural norms are communicated largely through unconscious cultural perspectives or 'stories'. Most of these 'stories' are implicit, unacknowledged and unquestioned. These 'stories' detail who we are, how we should live, our relationship to each other and the world, what constitutes as good and bad. Our industrialised cultural 'stories' allude that humans are the pinnacle of evolution, that humans need to be taught how to be 'good', that we are separate and dominant to nature, that the Earth is a limitless resource in service of our needs, civilisation is the best and only way to live, and that if we work harder and do more of what we did last year then it will be even better (Quinn 1992: 52–69). These 'stories' create individual and collective behaviours and at the same time justify existing ways of living. For example, if I comprehend the world as a resource, and the trees and water as reserves to be used in the service of my needs, it is appropriate, and predictable, that I seek ways to justify my (ab)use of those resources. It is also appropriate that I create conscious and unconscious emotional defences (Grille 2005: 210–12, Marmara 2005: 53–6, Hill 2009b: 50) such as denial and justification to protect myself from feeling the impacts of my actions. It is clear that the effectiveness of the current approaches to sustainability are limited and problematic, and that they are based on the 'stories' that have informed their creation. Fundamentally, unless these implicit 'stories' that dictate the functioning of industrial culture change, we will continue to create approaches and responses to sustainability based on these stories.

These stories that underpin industrial civilisation are designed around the misassumption that human nature is innately destructive (Marmara 2005: 53–6). In other words, a deep unconscious belief permeates dominant culture: that humans need to be socialised, modified, improved and constrained in order to be 'good'. This perspective is evident throughout society, in the education system, workplaces, religious institutions, prison systems, parenting methods and health care. For example, the education system assumes children require manipulation in order to learn, when learning is in fact a natural human trait (Neill 1953, Stallibrass 1989, Solter 1989, Hill 2003). Day in, day out for over fifteen years children are sent to school to be fed information that someone else has decided they need to know. They are tested against narrow standards of success, informed that they 'could do better' or 'have potential' if only they worked harder. If the child does not suit this regime they are labelled as problematic, poor performers (Posner 2009), excluded, reprimanded or in some way remediated through behaviour change strategies or medication to get them 'back on track'. These ways of relating are inherent across all systems in industrial culture and imply an inordinate lack of trust in human nature. When individuals experience mistrust it can engender conformity, rebellion and further their mistrust of others. All of which serve to disconnect individuals from their 'hearts and bodies and neighbours, from humanity and animality and embeddedness in the world we inhabit' (Jensen 2003: 41), bearing a high price on well-being individually and collectively.

These systems, such as the education system, are deeply destructive to human well-being. The systems we live in do not support and allow each individual's journey to unfold naturally. Quite simply, we cannot 'manifest who we really are' (Jensen 2003: 41). When our deep human need to be who we truly are is not acknowledged it can result in physical, psychological, emotional (Reich 1984:

37–9) and spiritual wounding and the creation of adaptive and maladaptive coping responses (Hill 2009b: 50). This repression of our natural constructive impulses (e.g. full emotional self-expression, spontaneity, creativity, connection with self/place/others) actually creates the current crises many are working so hard to overcome (Reich 1984: 37–9) and leads us to live in ways that are unsustaining personally, socially and ecologically. In other words, eco-societal problems like excessive consumption and commitment to progress are behaviours that can be associated with unhealed trauma caused by social systems that sever us from ourselves. As such, the aforementioned approaches and responses (Table 9.1) to sustainability are in fact features, not problems, of the cultural systems.

So where does all this take us? Fundamentally, a cultural transformation is necessary that is supported by new 'stories' that speak of our interconnectedness with the living Earth. This must include addressing the damage that industrial culture inflicts on human well-being by engaging in experiences that enhance our sense of self (Macy 1983: xiii) and heal cultural trauma. Otherwise we will continue to enact behaviour patterns that support and reproduce these destructive systems. The wonderful news is that this is already happening in pockets across the globe, predominantly as grassroots movements. There are many approaches that are engaging with personal and cultural transformation – some of these include permaculture, free-schooling, Transition Towns, Aware Parenting, somatic psychotherapy, homeopathy, deep ecology practices, 5Rhythms, social ecology, self-directed learning and a resurgence of indigenous perspectives. These approaches work with an understanding of the interrelationships influencing their area of focus. It's also evident from all of these initiatives that there is no blueprint for personal and cultural change. There are, in fact, many ways for approaching change from an integrated perspective. We can do things differently in our personal relationships, workplaces and schools and in the way we consume, spend time in nature, parent and express ourselves. It's essential that we are consistently critically self-reflective in these areas of our lives and that we are supported in doing so.

A social ecology approach to change can enable individuals to personally 'feel their own way' (Hill 1999: 197–201) to working with change in a way that is not driven by dogma, codes of behaviour or set theories of practice. Instead, the community of Social Ecology at the University of Western Sydney (2010) has supported many individuals to clarify their own values, perspectives, and experiences without the imposition of a particular agenda on what learning outcomes should be (Russell 1991: 126–9, Camden-Pratt 2008: 5–8). The high degree of congruence between what is taught and how it is taught creates a safe space for individuals to discover who they are. Social ecology uses processes that prompt individuals to explore their own meaning-making, personal experiences and actions in a broad range of contexts that can lead to transformative learning experiences (Taylor, 2001). Critiquing personal 'sets of values, beliefs, myths, explanations and justifications that appear self evidently true' (Brookfield in Taylor et al.: 2000: 81) can open up new possibilities and ways of experiencing, often leading to changes in approach to life. Thus, critical reflection (Friere 2005, Camden-Pratt 2008) plays a key role in social ecology. We have felt supported to develop our own way of working, and as a result, our approaches to working with change are different. Our diversity in approaches also creates fertile ground for engaging with the complexity of sustainability as we find that we can function in a much more creative, inclusive (of many perspectives and ways of knowing) and reflective modality. Moving away from notions of 'right' and 'wrong' we can better function in relationship with others and learn from one another. In our experience, approaching change in a way that is personally meaningful, sensitive to the current context (Kuhn 2009: 56–9) and values, the complexity of the 'system' (personal, cultural, environmental, unknown/spiritual) is much more likely to produce genuine change that is sustaining.

Our capacity for being sustainable is enriched through mutually supportive relationships. If we recognise that we live in an interconnected world, everything is in relationship (Heron 1999) and therefore is affected by and influences everything else. If our relationships are unsupportive, unloving and do not meet our needs it places great strain on our capacity to give support and behave in ways that are sustainable. The same is true for human–nature relationships. It's unlikely that our actions will be loving, considerate and sustainable towards nature if we do not have human relationships that enable and reflect these qualities. Many adults and children today have almost no relationship with the natural environment because they have had little or no direct experience of it (Louv 2008: 5–16). A lack of connection with nature 'leaves many people with a certain emptiness, no relationship, no grounds for thinking or caring about the environment' (Chenery 1994: 10). A disconnection with nature can spread to all our relationships; creating a sense of separateness with parts of others and ourselves, leading to the unconscious creation of compensatory behaviours (Hill 2009a: 6–7). There is significant value in connecting with local natural environments to fulfil our deeper human need for a 'sense of place' (Cameron 2003: 29–37). For example, in my (Jasmin Ball) work with young people in the outdoors I have been a witness to many transformations in individuals' perspectives of and relationships with nature. My approach seeks to enable young people to form their own unique relationships with nature and have meaningful experiences in their local environment. By facilitating a variety of ways of being in nature that can be fun, safe and engaging, individuals can feel deeply connected to themselves, their Earth and begin to nurture what they love and appreciate. My experiences suggest that it is essential we recognise how our relationships with self, others and nature influence ways of being in order to further enable us to live in sustaining ways.

In seeking a way forward, it may be helpful to reframe the term sustainability as 'sustaining'. This moves away from the notion that sustainability is a generalised, fixed state that is reached and labelled as such. Rather, by using 'sustaining', the concept can be reconsidered as an ongoing learning process that can invite important self-reflective questions such as: What is sustaining me? How can I better sustain life on the planet? What sustains my relationships? What sustains my workplace? What am I doing that does not sustain others, the planet and me? Where does my responsibility begin and end? How am I living my life in congruence with what is important for me? This can allow for a version of sustainability that is personally meaningful and contextually appropriate. Thus, 'sustainability' need not involve developing a consensual definition, rather, opening up a dialogue on personal meanings, values and vision for the world.

Closing reflections

Whilst it is essential that 'our task is transformational: of both our culture and ourselves' (Hill 2007: 1), our impact on the planet cannot continue to cause the current devastation while we realign with our deeper sense of self. Fundamentally, we need to find a balance between personal well-being, cultural re-storying and the impact of our lifestyle on the Earth. Essentially, these areas mutually influence and strengthen one another (Macy 1998: 17–23). However, over-reliance on one area within an individual person tends to lead to 'truncated, even dysfunctional' (Kasl and Yorks 2002: 5) perspectives. Consequently, personal wholeness, and therefore sustainability, is best fostered through integrating each of these aspects into our lives as well as in our wider society.

References

Beyond Blue (2010) *The National Depression Initiative*, http://www.beyondblue.org.au, accessed 5 March 2010.

Camden-Pratt, C. (2008) 'Social ecology and creative pedagogy: Using creative arts and critical thinking in co-creating and sustaining ecological learning webs in university pedagogies', *Transnational Curriculum Inquiry*, 5(1): 4–16, http://nitinat.library.ubc.ca/ojs/index.php/tci, accessed 10 September 2010.

Cameron, J. (2003) 'Dwelling in place, dwelling on earth', in John Cameron (ed.) *Changing Places: Re-imagining Australia,* Sydney, NSW: Longueville Media.

Chenery, M. (1994) 'Looking back from the bush: A view of eco-ethical thinking from the perspective of Australian Outdoor Education', paper presented at Art-Seminar on *Eco-ethical thinking in a cross-cultural perspective*, Germany: University of Saarland, July 1994.

Dittmar, H. (2008) *Consumer Culture, Identity, and Well-Being: The Search for the 'Good Life' and the 'Perfect Body'*, New York: Psychology Press.

Fien, J., Bentley, M. and Neil, C (2009) *People as Agents of Change: The Youth and Sustainable Consumption*, IYPF Working Group Australia, 20 January 2009, www.iypf.org, accessed 10 September 2010.

Friere, P. (2005) *Education for Critical Consciousness*, London & New York: Continuum.

Goff, S. (1997) 'A social ecology of participatory action research', *ALARPM Journal*, 2, 39–62.

Grille, R. (2005) *Parenting for a Peaceful World*, Sydney, Australia: Longueville Media.

Heron, J. (1999) 'The whole person as a web of relations', in *The Complete Facilitator's Handbook*?, London: Kogan Page Publishers.

Hill, S.B. (1999) 'Social ecology as future stories', in *Social Ecology Journal*, Volume 1, University of Western Sydney: School of Social Ecology and Lifelong Learning, 197–209.

Hill, S.B. (2003) 'Autonomy, mutualistic relationships, sense of place and conscious caring: A hopeful view of the present and future', in J.I. Cameron (ed.) *Changing Places: Re-imagining Australia*, Sydney, NSW: Longueville Media, 180–96.

Hill, S.B. (2007) 'Learning our way towards a sustainable and meaningful future', presentation at the University of Western Sydney, *Social Ecology Residential*, 10 July.

Hill, S.B. (2009a) 'Deep environmental leadership', in *Eingana*, The Victorian Association of Environmental Education, 32(2), 6–11.

Hill, S.B. (2009b) 'Enabling health and sustainability: From soil and landscapes to humans and cultures – a social ecology approach', paper presented at the *NRM Conference Changing Environments*, Geraldton, Australia, 29 October.

Hockachka, G. (2005) 'Developing sustainability: Sustainable development, not enough', in *Developing Sustainability, Developing the Self: An Integral Approach to International and Community Development*, Victoria, BC: Victoria University.

International Union for Conservation of Nature (2010) *Why is Biodiversity in Crisis?*, http://www.iucn.org/iyb/about/biodiversity_crisis/, accessed 19 April 2010.

Jensen, D. (2003) *Walking on Water: Reading, Writing and Revolution*, White River Junction, VT: Chelsea Green Publishing Company.

Kasl, E. and Yorks, L. (2002) 'An extended epistemology for transformative learning theory and its application through collaborative inquiry', in *Teacher's College Record*, http://www.tcrecord.org/content.asp?ContentID=10878, accessed 10 September 2010.

Komives, S.R., Lucas, N. and McMahon, T.R. (1998) *Exploring Leadership for College Students who Want to Make a Difference*, San Francisco, CA: Jossey-Bass.

Kuhn, L. (2009) *Adventures in Complexity: For Organizations Near the Edge of Chaos*, Axminster, UK: Triarchy Press.

Louv, R. (2008) *Last Child in the Woods: Saving Our Children from Nature-Deficit Disorder*, Chapel Hill, NC: Algonquin Books.

Macy, J. (1983) *Despair and Personal Power in the Nuclear Age*, Philadelphia, PA: New Society Publishers.

Macy, J. (1998) *Coming Back to Life: Practices to Reconnect Our Lives, Our World*, Philadelphia, PA: New Society Publishers.

Maiteny, P. (2009) 'Finding meaning without consuming: The ability to experience meaning, purpose and satisfaction through non-material wealth', in Stibbe, A. (ed.) *The Handbook of Sustainability Literacy*, Dartington, UK: Green Books.

Marmara, D. (2005) 'What do you do when words are not enough? The history and relevance of somatic psychotherapy', *Counselling Australia*, 5(5): 53–6.

Neill, A.S. (1953) *The Free Child*, London: Herbert Jenkins.

Peavey, F. (1994) 'Strategic questioning: An approach to creating personal and social change', in *By Life's Grace*, Philadelphia, PA: New Society Publishers.

Posner, R. (2009) *Lives of Passion, School of Hope: How One Public School Ignites a Lifelong Love of Learning*, Boulder, CO: Sentient Publications.

Quinn, D. (1992) *Ishmael: An Adventure of the Mind and Spirit*, New York: Bantam Books

Reich, W. (1984) *Children of the Future*, New York: Farrar, Straus and Giroux.

Russell, D. (1991) 'Social ecology in action, its rationale and scope for education and research', *Studies in Continuing Education*, 13(2): 126–38.

Solter, A. (1989) *Helping Young Children Flourish*, Goleta, CA: Shining Star Press.

Stallibrass, A. (1989) *Being Me and Also Us: Lessons from the Peckham Experiment*, Edinburgh: Scottish Academic Press.

Sydney Water (2010) 'Sustainability', *Sydney Water* http://www.sydneywater.com.au/Sustainability/, accessed 1 March 2010.

Taylor, E.W. (2001) 'Transformative learning theory: A neurobiological perspective of the role of emotions and unconscious ways of knowing', *International Journal of Lifelong Education*, 20(3): 218–36.

Taylor, K. et al. (2000) *Developing Adult Learners*, San Francisco, CA: Jossey-Bass.

Thurling, G. (2010) Personal communication with a social ecology graduate at University of Western Sydney, 5 March.

University of Western Sydney Handbook (2010) *Master of Education (Social Ecology)*, http://future.uws.edu.au/postgraduate_study/soc_sci/arts_social_ecology, accessed 1 March 2010.

Woog, R. and Dimitrov, J. (1994) 'Social ecology: A post-modernist, neo-positivist methodology', *Social Ecology Occasional Papers,* University of Western Sydney: School of Social Ecology and Lifelong Learning, 1: 1–5.

World Health Organization (2010) *Data and Statistics*, http://www.who.int/research/en/, accessed 21 July 2010.

10

A JOURNEY INTO PLACE

John Cameron

Teaching 'Sense of Place'

Joining the social ecology staff at UWS in the early 1990s gave me the freedom to explore my interest in the environmental significance of sense of place. In my previous work as a researcher and campaigner for the Australian Conservation Foundation (ACF), I had encountered people on both sides of the forestry conflict who expressed great care for the forests in which they lived and worked, but their voices went unheeded in the heated debate. They seemed to represent an attractive third force in the ritualised public conflict over logging, and I was moved and intrigued by their words. I despaired of making any meaningful progress as a protagonist, and gratefully accepted the opportunity to pursue environmental matters in an academic setting.

The School of Social Ecology created a more stimulating environment for social inquiry than I had ever experienced. In week-long residential schools, students were introduced to adult learning theory,[1] investigated their own preferred learning styles and designed projects for the semester accordingly. Staff members modelled this process by articulating the research questions they felt most passionate about and investigating them collaboratively.[2] In this supportive situation, I was encouraged to create my first course on sense of place.

Thus began a continually evolving dynamic between my own place relationships, student responses to the course design, my growing understanding of the place literature, and collaboration with my partner Vicki,[3] my colleagues and Aboriginal custodians.[4] At first I relied upon a three-day field trip into remote country as the setting for an exploration of the theme 'relationship with place', but students had difficulty relating these experiences with place theories or the environmental politics of place. When Vicki and I moved to a house with a large European garden and three acres of native forest adjoining the Blue Mountains National Park, I began gardening and regenerating the bush regularly. When working as an environmentalist, I had lived in rented accommodation or shared houses, and had paid far less attention to the local area than to the remote and rural parts of Australia in which I was researching and campaigning. I wanted to give students a taste of the power of affiliation with local places that had caught my interest at the ACF and that I was now beginning to experience for myself.

Social ecology emphasised the importance of the experiential learning cycle in which learners reflected on recent concrete experience, made sense of it conceptually and modified their subsequent actions accordingly.[5] I incorporated multiple learning cycles into the 'Sense of Place' course by requiring students to choose a place within half an hour's walk from their home, commit to visit it for at least two hours per week and record their impressions. Class members went through the fruitful exercise of determining the sorts of places in their local area that were most appropriate for their purpose in taking the course. The huge variety in the types of places selected – urban parks, degraded land, biodiverse bushlands, estuaries and forests – generated lively conversations between students about what they were doing each week in 'their place' and why.

The readings I first selected for the course were primarily North American place advocates[6] and European phenomenologists[7] or critics of place advocacy.[8] Students wanted more Australian readings, and in order to explore Australian place scholarship further, I invited three dozen academics, place writers and activists from around Australia to attend a place gathering. The culture within social ecology was very much to 'walk one's talk', so I designed an event in which the experience of the place itself (spacious sandstone overhangs in the deep canyons of the Upper Blue Mountains) was an integral part of the 'Sense of Place Colloquium'. Participants rose to the occasion, responding with poetry and passionate enthusiasm to the ideas of others, exploring why place matters, loving a place and grieving its destruction, what a politics of place might be, and the role of the university in promoting place awareness.

Inspired by the event, I modified my teaching practice and began reading aloud the words of poets and nature writers to students in the classroom and in the bush. I included newly discovered Australian place writing[9] in the course reader along with some of the colloquium papers. Encouraged by the ecopsychologists at the colloquium, I introduced the psychological dimensions of place into the curriculum. Initially resistant, the students soon began to make connections between their childhood special places and the type of place they selected as an adult. They could see that place had been, and was again, important to their understanding of their sense of self and path through life.

The second Sense of Place Colloquium on the theme of the interaction between Aboriginal and Western senses of place was held the next year in Central Australia, involving traditional Aboriginal custodians of the land along with local Landcare workers, teachers and tourist guides who worked with Aboriginal people. At the end of a week of walking with our guides and sitting in extraordinary country, we felt that there was an active non-human partner in the colloquium – the animate place itself. I returned to the classroom convinced of the centrality of Aboriginal understandings of 'country' for all Australians interested in place, and remodelled the 'Sense of Place' class. I added writings by and about Aboriginal people, augmented my class presentations and encouraged students to investigate the Aboriginal history of their places.

Vicki subsequently met a group of Aboriginal women artists from the community of Utopia, northeast of Alice Springs, who asked her to record their personal and Dreaming stories. She began to spend extensive periods of time with them at Utopia. As a result of her research[10] and the outcomes of the second colloquium, we designed a new subject, 'Place, Art and Culture in Central Australia' and took twenty-six students on a three-week field trip to the 'Red Centre' to experience the intercultural richness and complexities of place and art for themselves. With local guides, we camped on Aboriginal outstations, visited ancient rock carving sites and modern Aboriginal art galleries, walked through country with Aboriginal custodians, and witnessed the desperation and appalling conditions in which many Aboriginal people live. Vicki introduced the students to the daily practice of drawing and painting in their place journals and this turned out to be a skilful means of developing students' capacity for observation and reflective silence.

Many of the students wrote passionate and thoughtful accounts of what they had witnessed, and produced artwork that spoke powerfully of their experience of being in this 'country within a country'. Several of them subsequently moved to Central Australia to work with Aboriginal people. For our part, the next time we taught 'Sense of Place', we introduced painting and drawing practices as integral to students' weekly place visits. Bringing their place journals to class each week became a collaborative learning exercise as students could experience and be inspired by classmates' written and visual representations of their places.

In her own research into art and place, Vicki was confronting the major displacements in her own life as well as those suffered by the Aboriginal women with whom she was collaborating. Conversations between us turned to an acknowledgement of grief and loss of place, placelessness as well as sense of place. Because I had always felt so at home in the bush, I realised that I had not brought these topics into the classroom. As a result, the next time that Vicki and I co-taught the undergraduate 'Sense of Place' subject, we devoted a class to issues of placelessness and displacement. The response from the students was immediate. They talked about feeling alienated from where they lived, of losing childhood places they loved, of having no sense of place, and being previously ashamed to speak about it, as if not having a sense of place was a personal failure. The freedom to acknowledge alienation and loss allowed the students to approach their chosen places more honestly and realistically.

The third colloquium was held in a rural community north of Melbourne and focused on the implications of our sense of place work for life in the city. I recognised that I had to bring post-graduate students' experience of urban places more centrally into the class, and required them to spend a day in the most urban place that was within an hour's travel of their home. The exercise had a salutary effect on the students. Some had their expectations confounded by finding the central railway station more interesting than they could have imagined, or by discovering hidden 'secrets' in alleyways off busy streets. One student was so taken by his 'urban day' in Sydney that he ventured into the literature of the city and interwove his field notes with reflections on how James Joyce encountered Dublin. Even those students who had their negative preconceptions of the city reinforced were moved out of their comfort zone by experiencing it up close, and some went on to consider the social and environmental implications of a highly urbanised society.

While overseas on sabbatical leave, I was introduced to Goethe's way of science,[11] a method of responding to place in which natural phenomena reveal themselves to the sensitive observer through trained intuition. With the discipline of repeated drawing and visualisation practices, Goethean scientists develop a sufficiently receptive and intuitive space within themselves to allow the phenomenon to reveal its distinctive 'gesture' and essential nature. This offered a structured set of practices for students to develop a more participative relationship with their places. On my return, I introduced a lecture and field trip based on Goethean science, and received a wide range of responses. Some people clearly didn't grasp it, others were so moved they could barely speak, and quite a few integrated it into their understanding of the vast potential of place relationships.

Teaching 'Sense of Place' for fifteen years took me far beyond its original impetus. I have had the good fortune of working within the creative milieu of social ecology, with stimulating peers at place colloquia, having the guidance of Aboriginal custodians and a challenging and supportive co-teacher and life partner in Vicki. Many students found the course a deeply rewarding experience that transformed their view of the relationship between people and the places they inhabit, and the same may be said of me. Some followed their growing awareness of the environmental threats to their places into wider ecological consciousness, others became more reflective environmental advocates, while others have worked with farmers, social workers and corporations on place-related matters.

Changing places

After the fifth Sense of Place colloquium in southern Tasmania had successfully concluded, Vicki and I relaxed for several days with one of the co-organisers, Pete Hay, and his partner Anna Williams at their weekend 'shack' on Bruny Island. Several hours before we were due to fly back to Sydney, Pete took us on a pre-breakfast canoe excursion and proceeded to follow a white-faced heron further south than we had previously ventured. Just as we were becoming concerned about the time this was taking, the wide veranda of a simple house appeared in view, splendidly located on its own on the wooded shore. A few more strokes of the paddle brought us in sight of a For Sale sign on the shore. With half an hour to spare, we managed to meet the owner, had a very quick tour and shook hands on the sale.

We had not been looking for a holiday home, and certainly not one so far from the Blue Mountains. Over the following year we spent each university break at our new home on Bruny Island, and each visit confirmed the sense of belonging we felt when we were there. After a year, I took very early retirement and we sold our house in the Blue Mountains and moved to Bruny. We felt that in some mysterious way the heron had led us here and the place had called to us, and this feeling has deepened over the five years that we have lived at 'Blackstone'. Living a low-impact life and regenerating fifty-five acres of degraded land in a remote location has been the next step in our sense of place work.

Here we are fully immersed in the more-than-human world,[12] sharing our place with herons who nest in a nearby tree, shorebirds, wedge-tailed and sea eagles, woodland birds, wallabies, echidnas, possums and polka-dot quolls, and watching schools of fish, seals and pods of dolphins from our veranda. The briny smell of the Channel at low tide, the slap of the south-westerly gale on my face, the soft lap of the wavelets on the rocky shore, the play of last light on the ruffled waters – all these are a central part of a life without traffic, television, newspapers or near neighbours.

I've been able to devote more continuous time to Goethean scientific studies, setting off in the afternoons with my sketchbook and diary to sit quietly with some mushroom-shaped rocks on the tideline or a fire-scarred grasstree in the woodland. A sense of a 'conscious participation with Nature'[13] in which the land works on me as I work on it is slowly dawning, but it is far from the blissful union that these words might imply. The more I have opened myself up to life at 'Blackstone', the more conscious I have become of the mental habit patterns that get in my way, something that was less obvious in my busy life as a full-time senior lecturer. Watching the heron delicately pick its way along the edge of the water brings to life the qualities of poised attentiveness to which I aspire. Sitting regularly with the rock formations on the shore has become a lesson in inhabiting the various timescales at work in a place, from hundreds of millions of years to hours.

Once we became aware of the damage that had been done to our land by previous owners' over-clearing, ploughing and overgrazing, we embarked on a large land restoration project, planting over 4,000 local provenance native trees and native grasses. I spend every morning watering trees, maintaining tree guards and controlling weeds, which has given me abundant opportunities to practice the skills of receptive activity implicit in Goethean science. I am revisiting the literature on ecological restoration that I used to teach, especially the adage that to restore the Earth is to restore the self.[14] I do have a sense that we are working in partnership with the regenerative forces in the land, and we are consciously creating a sanctuary for wildlife here. I have relied on information from the occurrence of thistles, growth rates of trees and subsurface drainage patterns to guide me in the choice and placement of local tree species. Although it has been a physically revitalising process, from an ecopsychological point of view it has been as much about recognising my occasional adversarial cast of mind and how to persevere in the face of setbacks as restoring the self.

We now generate all our electricity through solar panels and a wind turbine, our rainwater tanks are our sole water supply, we have a vegetable garden and composting toilets. Our appreciation of the natural elements is not simply sensorial; at the same time as I feel invigorated by the gusting winds and blazing sun, I know our solar panels and wind turbine are charging the batteries for the house. Turning on the tap now connects us directly to our seasonal rainfall, and we no longer blithely assume there will be water or power available whenever we want it. The barriers to sustainability for me have been as much psychological as technological, but I'm coming to understand that acceptance of what we are given turns ecological limits into opportunities to learn and to flourish.

Our previous work on the interaction between Aboriginal and Western senses of place was soon brought home to us when we discovered that the partially hidden remains of an important intercultural historical site lay right next door to us. Through a series of uncanny events, we became the owners of that land and custodians of the site, and had it permanently listed on the Tasmanian Heritage Register. The knowledge that pivotal early contact between George Augustus Robinson and Truganini[15] took place on our land galvanised our imaginations and transformed Vicki's art practice. Reflecting on the role of this place and its inhabitants in bringing us here, showing us what is needed to be done on the land and guiding us to protect its heritage, I sense the power of 'country' as a sentient and active force in our lives. We had glimpses of this in Central Australia, but it is another matter to experience it in an ongoing way in one's home place.

Inevitably our journey has taken us beyond our own land into the island community. We are active members of the Killora Coastcare Group that is working to control invasive plants threatening the whole coastline. I was involved in the founding of the Bruny Island Environment Network (BIEN) in 2009, and have revisited issues of the environmental significance of sense of place. Bruny Island, like Tasmania as a whole, is deeply divided along environmental lines, especially forestry, so my usual approach to conflict, inclusiveness based on a mutual regard for place, has been challenged. I've taken a particular interest in how the Bruny community might best respond to global warming, and in workshops and talks I draw on people's lived experience of climate change rather than recapitulate the current highly polarised public debates.[16]

Writing my way through these encounters with place has become integral to my way of being here.[17] It is a means of making sense of the erratic, confusing and joyful experiences of life on Bruny Island, and it is a form of communication with the world beyond. Essay writing has been a way for me to illuminate the depth of my unfolding relationship with Vicki, the herons and all of the more-than-human inhabitants of 'Blackstone' and Bruny community members, and with the natural elements of the place. In company with Vicki's extraordinary sculptures, poems and paintings, it is a fertile expression of what that simple word 'place' has come to mean to us.

Notes

1 See Kolb (1984) and Brookfield (1986).
2 Staff were particularly influenced by models of cooperative inquiry as articulated by Peter Reason (1988).
3 Vicki (Dr. Victoria King) is a professional artist, writer and poet, and was formerly a senior lecturer in Fine Art at John Moore's University in England.
4 For a more detailed account of this process see Cameron (2008).

5 As discussed originally by Kolb (1984).
6 For example, Tuan (1974), Relph (1976), Hiss (1991), Snyder (1995) and Thomashow (1996).
7 Merleau-Ponty (1962) and Heidegger (1962), as well as the leading American place phenomenologist David Seamon (1993).
8 Foremost of whom was Massey (1994).
9 Such as Rolls (1984), Tacey (1995), Hay (2002) and Griffiths (1996).
10 Vicki included the Anmatyere and Alyawarre women's stories and her experiences at Utopia in her PhD, http://unsworks.unsw.edu.au/vital/access/manager/Repository/unsworks:778, accessed 7 February 2011
11 See Seamon and Zajonc (1998) and my detailed account of these discoveries in Cameron (2005).
12 This is David Abram's (1996) phrase that is more inclusive than terms such as 'non-human world' or 'nature'.
13 This is Henri Bortoft's (1996) phrase for one aim of Goethean science.
14 The subtitle to the first major text on ecopsychology by Roszak et al. (1995) is 'Restoring the earth, healing the mind'.
15 In 1828, Tasmanian Governor Arthur appointed George Augustus Robinson as 'Protector of Aborigines' to gather the remaining Aboriginals whose numbers had diminished alarmingly at the hands of European sealers, whalers and settlers. Robinson's so-called 'Friendly Mission' began at the 'sod hut' site on our land when he met with thirteen Nuenone people including Truganini, daughter of the leader of the Nuenone people. She subsequently accompanied Robinson as guide and translator on his trek around Tasmania that ended in exile for her people on Flinders Island.
16 I refer here both to our own experience of reducing our ecological footprint as well as to the prominent place scholar Edward Relph (2009) who advocates a phenomenological approach to addressing global warming.
17 A series of my essays on our experiences at Blackstone is being published in *Environmental and Architectural Phenomenology*, edited by David Seamon. It is available online at http://www.arch.ksu.edu/seamon/EAP.html.

References

Abram, D. (1996) *The Spell of the Sensuous*, New York: Pantheon Books.

Bortoft, H. (1996) *The Wholeness of Nature: Goethe's Way Towards a Science of Conscious Participation in Nature*, New York: Lindisfarne Press.

Brookfield, S. (1986) *Understanding and Facilitating Adult Learning*, San Francisco, CA: Jossey-Bass.

Cameron, J. (2005) 'Place, Goethe and phenomenology: A theoretic journey', *Janus Head*, 8(1): 174–98.

Cameron, J. (2008) 'Learning country: A case study of Australian place-responsive education', in D. Gruenwald and G. Smith (eds) *Place Based Education in the Global Age: Local Diversity*, New York: Lawrence Erlbaum.

Griffiths, T. (1996) *Hunters and Collectors: The Antiquarian Imagination in Australia*, New York: Cambridge University Press.

Hay, P. (2002) *Vandiemonian Essays*, Hobart: Walleah Press.

Heidegger, M. (1962) *Being and Time*, New York; Harper & Row.

Hiss, T. (1991) *The Experience of Place*, New York: Vintage Books.

Kolb, D. (1984) *Experiential Learning: Experience as a Source of Learning and Development*, Englewood Cliffs, NJ: Prentice Hall.

Massey, D. (1994) *Space, Place and Gender*, Cambridge: Polity Press.
Merleau-Ponty, M. (1962) *Phenomenology of Perception*, London; Routledge.
Reason, P. (ed.) (1988) *Human Inquiry in Action: Developments in New Paradigm Research*, London: Sage.
Relph, E. (1976) *Place and Placelessness*, London: Pion.
Relph, E. (2009) 'A pragmatic sense of place', *Environmental and Architectural Phenomenology*, 20(3): 24–31.
Roszak, T., Gomes, M. and Kanner, A. (eds) (1995) *Ecopsychology: Restoring the Earth, Healing the Mind,* San Francisco, CA: Sierra Club Books.
Seamon, D. (ed.) (1993) *Dwelling, Seeing and Designing: Toward a Phenomenological Ecology*, Albany, NY: State University of New York Press.
Seamon, D. and Zajonc, A. (eds) (1998) *Goethe's Way of Science: A Phenomenology of Nature*, Albany, NY: State University of New York Press.
Snyder, G. (1995) *A Place in Space: Ethics, Aesthetics, and Watersheds*, Washington, DC: Counterpoint.
Tacey, D. (1995) *Edge of the Sacred: Transformation in Australia*, Melbourne: HarperCollins
Thomashow, M. (1996) *Ecological Identity*, Cambridge, MA: MIT Press.
Tuan, Y.-F. (1974) *Topophilia: A Study of Environmental Perception, Attitudes and Values*, Englewood Cliffs, NJ: Prentice-Hall.

Further reading

King, V. (2005) *Art of Place and Displacement: Embodied Perception and the Haptic Ground*, PhD, University of New South Wales, Sydney, http://unsworks.unsw.edu.au/vital/access/manager/Repository/unsworks:778, accessed 7 February 2011.
Rolls, E. (1984) *A million Wild Acres*, Melbourne: Penguin.

11

THINKING AND ACTING LOCALLY AND GLOBALLY

Martin Mulligan

Feet to the ground in social ecology

Joining the staff in the Social Ecology Program at the University of Western Sydney (UWS) in 1993 was a very grounding experience for me. First there was the historic and atmospheric campus that I had visited numerous times before as a conference attendee. Not far away was the Hawkesbury River where I had played as a child while my father and his friends were water-skiing. Beyond that, as a backdrop to the campus, were the Blue Mountains where I had spent countless weekends bushwalking when I was a teenager. There was both a familiarity and a tantalising mystery about the place where I would be working. Secondly, the experiential learning paradigm – which underpinned the development of the Social Ecology Program – made me think quite deeply about why I had applied for this job. The focus on 'lived experience' was sometimes unsettling yet it was also timely in this transitional phase in my life. In some ways it felt as though I had found a home – a 'safe refuge' – and yet I sensed that I was in for some 'edgy' and transformational experiences.

After a few years it seemed to me – and others on staff – that the heavy emphasis on learning processes meant that we did not absorb enough ideas and insights from outside our own experiences. We wanted a bit more 'content', as well as process, in our teaching and learning. I watched with great interest as John Cameron developed his groundbreaking post-graduate course on 'Sense of Place' – drawing on the work of cultural geographers, human geographers, philosophers and creative writers – and I developed an introduction to this work in an undergraduate course I was teaching. What John taught, and I absorbed, was that there are techniques for increasing our sensitivity to, and knowledge of, the myriad places in which we dwell. We can consciously develop strategies for becoming more 'place responsive', to use John's term. This opened the door for a better understanding of the 'phenomenological' tradition that we had long espoused by recommending books such as Max Van Manen's (1990) *Researching Lived Experience.* It also introduced me to the work of UK-based human geographer Doreen Massey and her important warnings against 'place essentialism' (1994). A romantic, parochial, interpretation of place identities could deepen social divisions within class- and gender-divided communities that had also become 'multicultural' through processes of global migration, Massey noted.

From the local to the global and back again

After leaving social ecology to migrate to Melbourne – in order to be closer to my wife's extended Sri Lankan family – I happened upon a short-term position in the rather grandly named Globalism Institute at the Royal Melbourne Institute of Technology University (RMIT). Researchers within the Institute had already initiated a big research project on the 'well-being' of local communities across Victoria with the Victorian health promotion agency, VicHealth, and I got a small grant to explore ways to introduce place awareness into this study, within the case study communities centred on outer-urban Broadmeadows and the popular central highlands town of Daylesford. Globalism Institute Director Paul James was interested in what a phenomenological approach might add to a research design that was primarily based on a conception he had developed about interactions between 'layers' of society, ranging from 'face-to-face communities' to more spatially extended social formations, such as national and global 'imagined communities'.[1] A globalising tendency across human history, James has argued (2006), has not so much replaced older forms of social integration – such as tribalism – but rather layered new forms on top and it has now become important to understand how the global resides within the local and vice versa. Quite unexpectedly, an opportunity had opened up for me to work on people–place relationships within several Victorian local communities and yet to do so within a local–global framework.

As my position in the Globalism Institute became more secure I took charge of the large study that the Institute was conducting for VicHealth on the contribution that community arts and celebrations can make to the well-being of people living in local communities (mentioned above). The fieldwork for this study was carried out across a spectrum of local communities in Victoria, ranging from inner urban St Kilda, to the aforementioned Broadmeadows and Daylesford and including the regional community centred on the western Victorian town of Hamilton. The result was a 177-page report titled *Creating Community: Celebrations, Arts and Wellbeing Within and Across Local Communities* (Mulligan et al. 2006), which circulated widely across Australia. This was followed by a study conducted for the Australia Council for the Arts on ways in which community art work could enhance local governance and this resulted in a report, titled *Art, Governance and the Turn to Community* (Mulligan and Smith 2010), that was launched publicly in 2010. These two reports argue that well designed and adequately resourced community art projects and programs can play a major role in creating a more inclusive sense of community at local levels in a world of great uncertainties. As the title of the first reports suggests, we are arguing that a sense of community is not a 'given' in the 'liquid' conditions of late modernity. Furthermore, a narrow conception of a community's identity can create tensions, divisions and even conflict within complex, multilayered, local communities. An inclusive sense of community has to be created and constantly recreated in response to changes that might be taking place within the community itself or in response to changing conditions or influences that might be emerging at national or global levels. Participatory art has a significant role to play here because it can help individuals to 'make their experiences cohere' by creating a sense of 'narrative movement' about their lives, as sociologist Richard Sennett (2006) put it, and when they are working collectively with other people within their local communities on community art projects individuals can gain a sense of how their personal narratives fit into broader, collective, narratives.

The 'wilful construction' of inclusive communities

Our research has supported the conception of community in Gerard Delanty's excellent book of 2003, i.e. that community should not be seen as a social structure but rather as a normative sense of belonging that must be 'wilfully constructed'. Whereas many sociologists and cultural theorists – ranging from Emile Durkheim to Eric Hobsbawn and Iris Marion Young – have argued that the notion of community has become less relevant in the conditions of 'modern' life, there is plenty of evidence to suggest that the desire for community has actually grown stronger and Delanty explains this by saying: 'Community is relevant today because, on the one hand, the fragmentation of society has provoked a worldwide search for community and on the other … cultural developments and global forms of communication have facilitated the construction of community' (2003: 193).

Near the end of his book, Delanty noted that the 'revival of community is undoubtedly related to the crisis of belonging in relation to place' (ibid.: 195). In our research for VicHealth we were able to demonstrate that 'sense of place' work could indeed enhance a collective sense of local belonging. However, we also draw from the work of Zygmunt Bauman (e.g. 2001) in noting that global flows of people and ideas will constantly disrupt a local sense of belonging and concur with Doreen Massey (1999 and 2005) in saying that claims about the essential identity of local communities must take into account the 'coexisting multiciplicity' of contemporary local communities. There is a big difference, of course, between a sense of community and an *inclusive* sense of community and this is where the 'social inclusion' rhetoric that the Rudd Labour government in Australia inherited from the Blair Labour government in the UK could be given more substance. Our research for the Australia Council for the Arts certainly confirmed that there is a critical role for local government to play in the creation of an inclusive sense of local community in a world of great change and uncertainty. Indeed it can be argued that this should be a central goal of local government. As UK social theorist Nikolas has put it, a more globally integrated world means that 'community is not simply the territory of government but a means of government' (2008: 93).

In 2009, the Globalism Institute at RMIT morphed into a school-based Globalism Research Centre (GRC) when a university-wide Global Cities Research Institute came into being and, in 2010, I became the director of the GRC. With generous support from a local benefactor we have been able to continue working with the regional community in the Hamilton district of Victoria where we were asked to help local community activists think through the possible impacts of climate change and other global challenges, such as 'peak oil'. We introduced an innovative form of 'scenarios mapping' to create some plausible yet challenging stories about what life might be like in the region in 30–50 years time and we have worked with community leaders in the town of Coleraine to create a 'Coleraine Enterprise', which is looking at practical action in areas such as water recycling, the diversification of food production and greater use of solar power in the town. We are also working on a history of the Hamilton region – titled *Whatever Happened to Australia Felix?* – which aims to show that global influences have shaped local history ever since the first European settlers arrived in the 1830s.

Turning to community in times of crisis

Outside Australia, GRC is conducting research on interrelationships between the local and the global in a wide range of local communities in Papua New Guinea, East Timor, Malaysia, Sri Lanka and India. For me personally, the biggest project by far has been a four-year study on the rebuilding of local communities following the 2004 tsunami disaster in Sri Lanka and India. Fieldwork for this study was conducted in two communities in southern Sri Lanka, two communities in eastern Sri Lanka (which had also been badly affected by the Sri Lankan civil war), and with people who were relocated away from eight coastal villages in northern Chennai into two new permanent settlements away from the sea. Sadly, this study confirmed that a lot of post-disaster aid funding was wasted and in some cases the poor delivery of aid created new divisions within local communities and made 'vulnerable' people – such as single parents or those with few job skills – even more vulnerable. Inequitable delivery of aid has made ethnic divisions in eastern Sri Lanka even worse and has created new grounds for resentment among those living in the east and north of the island. Fortunately, we were able to find some examples of good practice in regard to the building of inclusive and resilient post-disaster communities and we were able to analyse some good and effective partnerships between local organisations and community activists and international aid agencies. Local knowledge was often underrated in the post-tsunami recovery work and the good practice that we were able to observe and document involved a relatively early shift from relief to longer term social planning. As one effective aid worker put it,[2] the aim should not have been to simply put the disaster survivors into houses plonked into empty paddocks but rather to imagine how the resettled communities might be working in 10 or 20 years time. Our report for AusAID[3] recommends that a more 'deliberative' and far-sighted approach must be taken in moving from disaster relief to the rebuilding of resilient and inclusive communities. The post-tsunami research confirms that an inclusive sense of community must be 'wilfully constructed' in the contemporary world. This holds for Sri Lanka as much as it does for Australia.

The tsunami disaster was a particular kind of 'external' challenge for unsuspecting local communities in Sri Lanka, as well as Indonesia, Thailand and India. In 2008 Sri Lanka also had to grapple with a sharp increase in food prices, due to the surge in global oil prices, and for the first time Sri Lanka had to import rice from India to cope with food shortages. Ironically, the onset of the global financial crisis (GFC) in October 2008 forced down global oil prices and took the heat out of rising food prices. Yet the GFC demonstrated that there are ongoing structural weaknesses in global financial practices and institutions and the easing of the global recession has resulted in a new upsurge in oil prices. Countries such as Sri Lanka have barely contemplated the challenges of climate change and yet this shapes as one of many 'global challenges' that the country will face in the years ahead. The Copenhagen summit on climate change, at the end of 2009, made some progress in building a global consensus that climate change is a real and present threat to humanity even if the representatives of powerful national governments were not able to lift their eyes high enough to negotiate a meaningful framework for global action.

Climate change as a local and global challenge

David Marr has made it clear (2010) that Kevin Rudd took the setbacks in Copenhagen very badly because he had staked his own reputation on a successful outcome. Under pressure from opposition political parties led by globally recalcitrant climate change sceptics, Prime Minister Rudd announced a humiliating backdown on the central plank of the government's climate change response – the implementation of an Emissions Trading Scheme (ETS). Subsequent public opinion polls demonstrated that this led to a sharp decline in public confidence in Rudd and his government because they were left without a strategy for tackling what Rudd himself had called the 'greatest moral challenge of our time' and, indeed, contributed significantly to the close election result of 2010. Part of the problem was that the Rudd government had focused much too heavily on an ETS as *the* response to climate change when it could be, at best, a very partial response.

Advocates of 'cap and trade' emissions trading schemes argue that they are superior to a straight out carbon tax because they give affected industries and enterprises the opportunity to use market mechanisms to adjust to low carbon economies. There may be some truth in this and market mechanisms for achieving 'structural adjustments' should not be ruled out. However, market mechanisms can also result in a relocation of the problem to somewhere else (perhaps outside Australia where environmental regulations may be more lax). They can allow space for the continuation of greenhouse gas pollution at a local level – as in the continuation of coal-fired electricity generation in the Latrobe Valley in Victoria – which has both local and global consequences. A national emissions trading scheme will do nothing to prevent Australia sending shiploads of coal to China as fast as it can. As George Monbiot (2006) has effectively argued, we cannot simply rely on a free market economy – which has clearly exacerbated the problem of global climate change – to fix itself. There is clearly a need for more stringent regulation – nationally and globally – to reduce greenhouse gas emissions significantly and, as Monbiot also argues, there is an urgent need to think beyond 'economic policy' in order to bring about significant changes in social attitudes and practices.

While the Rudd government has focused so narrowly on its proposed ETS, there has been a surge in local activity aimed at adapting to climate change at a local level.[4] Local action is certainly important and that is why the GRC is supporting the Coleraine Enterprise, mentioned earlier. However, local action should not be counterposed to action taken at regional, national or even global levels. If ever a challenge posed the need to think and act locally and globally at the same time it is the challenge of climate change.

In his book *Bringing the Biosphere Home,* Mitchell Thomashow (2002) points out that we connect with extensive natural systems and cycles – such as the hydrological cycle – locally and in an embodied way. We can use ecological knowledge and our personal senses to gain a much deeper understanding of our participation in natural cycles and rhythms that are being altered by global climate change. If we connect Thomashow's starting point with James Lovelock's (2000) way of thinking about the planet and all its systems as a single living system – *Gaia* – then we can use sense of place work at a local level to enter into a much stronger global awareness. In the past, local environmental conservation campaigns have often focused on the preservation of local ecosystems and 'endangered species' and this will continue to be an important starting point. However, climate change renders any form of parochialism dangerous and irrelevant. Climate change sharpens the need to understand the interconnection of the local and the global and it sharpens the ethical need to think and act locally and globally at the same time.

Climate change does, indeed, pose the greatest moral challenge of our time. We need to reinvent local life, ensure that our nation provides practical and moral leadership on responding to the challenge, and acknowledge that what we do in Australia has an impact on every person and every living thing on this planet. At the same time, the 'revival of community' in times of great uncertainty offers the hope of achieving a much better balance between the individual and the social. As Gerard Delanty has suggested, efforts to create more inclusive and purposeful communities could lead to a 're-enchantment of everyday life' (2003: 132).

Notes

1 The notion of 'imagined communities' comes from Benedict Anderson's book titled *Imaginary Communities* (1983).
2 Philippe Fabry, who founded a post-tsunami NGO called Sri Lanka Solidarity.
3 The report can be accessed through the GRC website at www.rmit.edu.au/globalism.
4 As seen in the growth of the Transition Towns movement in Australia and internationally.

References

Anderson, Benedict (1983) *Imaginary Communities: Reflections on the Origin and Spread of Nationalism*, London: Routledge.

Bauman, Zygmunt (2001) *Community: Seeking Safety in an Insecure World*, Cambridge, UK: Polity Press.

Delanty, Gerard (2003) *Community*, London and New York: Routledge.

James, Paul (2006) *Globalism, Nationalism, Tribalism: Bringing Theory Back In*, London: Sage Publications.

Lovelock, James (2000) *Gaia, a New Look at Life*, Oxford: Oxford University Press.

Marr, David (2010) 'Power trip: The political journey of Kevin Rudd', *Quarterly Essay*, June 2010, Melbourne: Black Inc.

Massey, Doreen (1994) *Space, Place and Gender*, Cambridge, UK: Polity Press.

Massey, Doreen (1999) 'Spaces for politics', in Doreen Massey, John Allen and Philip Sarre (eds) *Human Geography Today*, Cambridge, UK: Polity Press.

Massey, Doreen (2005) *For Space*, London: Sage Publications.

Monbiot, George (2006) *Heat: How to Stop the World From Burning*, London: Penguin Books.

Mulligan, Martin and Pia Smith (2010) *Art, Governance and the Turn to Community: Putting Art at the Heart of Local Government*, Melbourne: Globalism Research Centre, RMIT University.

Mulligan, Martin, Kim Humphery, Paul James, Christopher Scanlon, Pia Smith and Nicky Welch (2006) *Creating Community: Celebrations, Arts and Wellbeing Within and Across Local Communities*, Melbourne: Globalism Research Centre, RMIT University.

Rose, Nikolas (2008) 'The death of the social: Reconfiguring the territory of government', in Peter Miller and Nikolas Rose (eds) *Governing the Present: Administering Economic, Social and Personal Life*, Cambridge, UK: Polity Press.

Sennett, Richard (2006) *The Culture of New Capitalism*, New Haven, CT: Yale University Press.

Thomashow, Mitchell (2002) *Bringing the Biosphere Home: Learning to Perceive Global Environmental Change*, Cambridge, MA: The MIT Press.

Van Manen, Max (1990) *Researching Lived Experience*, London and Ontario: State University of New York Press.

12

THE POWER AND INFLUENCE OF THE SYNTHETIC CORTEX

Bruce Fell

Easy Terms and Conditions!
Everything Must Go
Buy Now
Pay Later
SALE!

I can't remember a time when there wasn't an advertisement informing me about the latest sale. Such placards and banners wallpaper my world, constantly reminding me to buy now and pay later. These are strange reminders and strange memories; strange, because at first they don't seem like memories.

Memory per se is directly linked to how we neurologically process the world; how our priorities and attitudes surrounding self, sexuality, religion, design, politics and ecology are formed, reinforced and passed on to the next generation. Today we are entwined in a *Buy Now Pay Later* corporate memory: it is how we are constantly reminded to think. To this end, the product wrappers we interact with (Internet, television, billboards, newspapers, etc.) are more than vehicles for product promotion; they are Memory with *a thousand faces* (Campbell 1972), they reiterate, again and again and again the memories we are reminded to remember.

This chapter is all about memory: my memory, your memory and, importantly, our collective memory. Understanding how memory works helps us understand how tribe, city and civilisation work. Neuroscience, in combination with cognitive archaeology, informs us that memory is located in two places: internally and externally. This chapter is about this relationship. Technology per se is the depository of external memory; in essence, the two can't be divided (Heidegger 1977). The following example illustrates the relationship between external and internal memory; a blogger is musing about their relationship with an iPhone:

> A whole lot of my cognitive activities and my brain functions have now been uploaded into my iPhone. It stores a whole lot of my beliefs, phone numbers, addresses, whatever. It acts as my memory for these things. It's always there when I need it … I have a list of all of my favourite dishes at the restaurant we go to all the time in Canberra. I say, OK, what are we going to order? Well,

> I'll pull up the iPhone – these are the dishes we like here. It's the repository of my desires, my plans. There's a calendar, there's an iPhone calculator, and so on. It's even got a little decision maker that comes up, yes or no.
>
> (Lehrer, 2009)

The iPhone example goes some way towards highlighting the symbiotic relationship between external and internal memory: influencing one, influences the other. There are numerous examples of this across a wide spectrum, for example the selection criteria deemed important by custodians of imperial memory at museums, libraries and media outlets impact on our decision making. The films, DVDs, books, tools and works of art selected for preservation shape collective memory – increasingly, multinational corporations finance the production and preservation of popular external memory.

External memory constantly reminds us of the dominant discourse of the day: the memories those with power deem important. This is how external memory has functioned throughout human evolution, most noticeably over the past 50,000 years. As in the past, external memory impacts on human well-being. Today, by and large, the memories resulting from what we Google, click, switch, play and unwrap are contributing to global ecological degradation.

Mainstream memory needs an upgrade.

Arguably, the memory system that underscores our day-to-day lives hasn't had a major upgrade since 1944. This is when the International Monetary Fund (IMF) and World Bank were developed at a major economic conference at Bretton Woods, USA. At the opening session of the conference a welcoming message from President Roosevelt said the new system would function as: 'a dynamic world economy … on an earth infinitely blessed with natural resources … the peoples of every nation will … enjoy increasingly the fruits of material progress … prosperity has no fixed limits' (Coward and Maguire 2000: 29; also see Lipsitz 1990).

The 1944 Bretton Woods system equated economic growth with human well-being; it didn't link human well-being to ecological sustainability. Subsequent minor upgrades have enabled this economic centric system to run faster, which doesn't mean, in ecological terms, better. It has long been argued by those opposed to the current system, that civilisation is running on an outdated operating system; that unless we develop the memory to run an ecologically sustainable system, we will crash (Fell 2005).

When a civilisation loses the ability to upgrade its memory, it eventually becomes hamstrung by its old memory: its old attitudes. For example, old memory computed the world as flat, that slavery was legitimate, that women couldn't vote, that indigenous persons were inferior. The old memory of the current system computes economic development over biodiversity, household consumption over quality of life; it seems unable to upload the new memory required to run *human well-being and nature* software.

If we stay with the computer metaphor, we can recognise that the function of a computer's brain, its CPU (central processing unit), activates internal and external memory as if they were coming from the one portal. It's only when we disconnect the USB (or Wi-Fi) that the computer is unable to access all the memory it requires. As subsequent memory upgrades pass the system by, the output from the CPU becomes outmoded. Over time, activists and progressive leaders have found various ways to access the governing CPU in order to upgrade much of its old data. As a result, our system now computes that: 'All human beings are born free and equal in dignity and rights' (Universal Declaration of Human Rights). The time has come for another upgrade, the current 'Easy Terms and Conditions! / Everything Must Go / Buy Now Pay Later' system exemplifies the type of low-grade memory that our current system is churning out. How to upgrade the system, to add new memory, is the challenge.

Our mind is the portal through which all upgrades are installed.

Just as a USB port or Wi-Fi portal enables external memory to interface with internal memory, the human Mind can be thought of as the portal linking internal and external memory (see Broks 2003, for a discussion of Mind). The origins of external memory are found in the mutation that brought into existence our large brain, and, by association, the human mind (Lynch and Granger 2008). Gifted with a large cranium packed to the hilt with cortex, people began to create complex external memory (Hodgson 2000). From prehistoric images painted on the walls of caves to the current messages posted on Facebook walls, we have developed a vast catalogue of external memory by combining art, craft and code (alphabet, etc.) onto a variety of surfaces: rock, wood, canvas, film, LCD screens, etc. Just as we should not underestimate the consequences of having a large brain, we should also not underestimate our nimble body; our dexterity allows us to render onto an unlimited variety of services complex thoughts created by our internal cortex. No other sentient being has this unique brain–mind–body ability to affix memory onto rock, wood, fibre, iPhone, chocolate wrapper, etc. Humans have due to the capacity of their internal cortex in combination with their nimble body created a 'synthetic cortex' – an external means for storing memory.

The synthetic cortex is the defining apparatus created by the human mind.

The rest of this chapter will discuss why it is vital, in terms of human well-being and ecological sustainability, to understand the power and influence of the synthetic cortex. To fully appreciate the influence of the synthetic cortex we need to consider human brain development. The first marks made by a young person upon a surface are similar in nature to the marks made by other sizeable sentient beings. Such inconsequential impediments made during the process of achieving some other goal have no purpose. For instance, the marks made when a child smudges food across the tray of its highchair, like a swath of blood brushed across the earth as a wolf carries its prey to a den, are inconsequential marks. Such rough looking calligraphy is caused by unintentional mark-making. Yet, there comes a time when our growing brain reaches a capacity that enables it, in fact demands that our body make marks (see Dissanayake 1992). For example, all healthy persons make the same shaped marks, in the same order, at the same stage of cognitive development: line, circle, triangle and square. As we begin to make marks, we also begin to become aware of marks made by others: television, Internet, billboards, etc. As this complex process begins, we read and write, write and read the synthetic cortex *prior to our ability to fully contextualise our actions* – we interact with the synthetic cortex prior to being mindful of the fact. This is how the synthetic cortex influences our foundational mind, for it communicates to us prior to our being cognisant of constructed and orchestrated communication (Fell 2009: 183–233).

With repetition our rough neural 'writing and reading' tracks become smooth symbiotic pathways as our mind returns again and again to a vast array of external memory. As we interact with the technologies of art, craft and code our mind has, without knowing it, escaped from the confines of our internal cortex. Now, until death, our mind will oscillate between internal and synthetic cortex.

As we have seen, due to the size of our brain and the dexterity of our body we create external memory upon the synthetic cortex as a consequence of being human, as all human societies have done, always (Bednarik 1990). The question arises: 'whose memory dominates the contemporary synthetic cortex?' The answer to this question provides an insight into the current state of human well-being and global ecological sustainability.

When a person dies (mathematician, artist, wheelwright, etc.) the memory of their code, art and craft lives on. Memories surrounding worship, design, seasonal planting, sexuality, herd migration, biodiversity, etc. are passed on via the synthetic cortex from one generation to the next. When the internal memory of the mathematician, artist, wheelwright, etc. is uploaded onto the synthetic cortex it becomes available, potentially, to all. While the synthetic cortex has substantially expanded over the past 50,000 years, the average size of the human brain hasn't changed in that time. At birth, the brain dimensions of our predecessors had the same form as our own contemporary internal cortex. Arguably, the central difference between 'them' and 'us' is the type, capacity and accessibility of the available external memory devices: the synthetic cortex. Each generation has the potential to access more sophisticated external memory than the previous generation: from rock art to iPad. The question arises: does sophisticated memory equal 'better' memory? Arguably, a sustainable future resides in the answer to this question. By 'better memory', I mean memory that links human well-being with global ecological sustainability.

As mentioned, the ability to store memory beyond the life of an individual underscores the power and influence of the synthetic cortex. Internet, television, film, radio, newspapers, billboards, books, theatre and galleries underscore the contemporary memories we access. They also erase memory. Memory is fragile, internal memory disappears at death, or earlier in the case of dementia (Greenfield 2008) or stress (Sternberg 2009). External memory disappears due to the defacement, erasure and re-rendering of the synthetic cortex. If the tendril between internal and external memory is severed, individual and community memory dissipates. There are numerous examples throughout history where invaders and emerging powerbrokers have torn down temples, burnt books and generally defaced the memory of an existing culture in order to impose new memories on the synthetic cortex – from which they hope a new culture will emerge, one in which 'their' memories will become 'our' memories. One consequence of this is that traditional culture and biodiversity has been, and continues to be, placed in danger.

Despite the sophistication of the current technologies that disseminate memory, mainstream society has forgotten about the link between human well-being and ecological sustainability: one need only walk through a typical suburban estate, shopping mall or industrial precinct to appreciate the extent of the memory loss. How did ecologically sustainable memories, held by indigenous peoples the world over, become erased from the contemporary synthetic cortex? In part this came about as a consequence of the memories that contemporary powerbrokers deem important; the memories they deem worth remembering. As such, popular memory is devoid of the memories that our powerbrokers deem unimportant. Memories deemed unimportant are erased – if only by the fact that they are not preserved and promoted. Aspects of indigenous Australian memory spanning 50,000 years or more were lost when Europeans vandalised the synthetic cortex of indigenous culture. Both the violent and passive destruction of the synthetic cortex affects our day-to-day memory. Throughout the ages those with power (shaman, general or monarch; artist, state or corporation) have dominated the synthetic cortex, be that local, national and, now, global.

The central *Buy Now Pay Later* tenet of the Bretton Woods memory system has become a pervasive internal and external memory in the day-to-day lives of contemporary sophisticated persons. Arguably, the intention of the Bretton Woods system was honourable, it set out to facilitate world peace via economic stability in response to a debilitating global depression sandwiched between two world wars. In the end, Bretton Woods, like its offshoots, was based on a mistaken premise. The Earth is not 'infinitely blessed with natural resources', as a result the peoples of every nation *will not* 'enjoy increasingly the fruits of material progress'. The reality is, as the natural and social sciences have demonstrated, prosperity has very real 'fixed limits'.

The commercial corporate memory rendered upon the contemporary synthetic cortex does not encourage us to think and act in a sustainable manner. From the moment each contemporary individual, as a child, begins absorbing the dictums of the synthetic cortex, unbeknown to them, they begin remembering a mistaken understanding of the relationship between human well-being and global ecological sustainability. The proof is in the pudding: a child born in 2010 faces a landscape far more degraded and with less biodiversity than a child born ten years ago. Will this change by 2020? Our unquestioned assumptions and deep neurological acceptance of the outdated stories currently rendered on the synthetic cortex undermines global ecological sustainability.

How do we move forward?

Returning to the computer metaphor, just as one can manipulate certain sites to appear first in a Google search result, certain memories can be placed at the forefront of our mind. Today, not unlike lollies and magazines stacked around a supermarket checkout, the sweet taste and bright cloth of the prevailing myth sweetens the sensation and colourfully covers over the exploitation of other people, places and species. Ways to educate and prevent society from being seduced by such unethical promotion requires attention. The challenge being that it is nearly impossible, neurologically, to comprehend what is not represented on, or erased from, the synthetic cortex. For example, in 1970, Joni Mitchell's ecological anthem sang, 'you don't know what you've got until it's gone'. Forty years on, one can be excused for thinking the 'parking lot' that Mitchell derided is natural. In this sense, the ability of the synthetic cortex to erase memory presents educators and reformers with a twofold challenge – the trees, grass, birds, streams and pleasure erased, in order to build the 'parking lot', requires physical re-creation, let alone the re-rendering of the symbols of sustainability onto the synthetic cortex.

Within the twofold challenge we need to investigate unquestioned assumptions concerning new technology (iPad, etc.), because new technology tends to draw on old memory, as opposed to creating 'new' memory. For example, new technology does not contain the miraculous ability to automatically remember the trees that grew where the parking lot now stands! The iPhone example above remembers the same restaurant menu as the waiters' printed version. In this sense, we should not equate 'speed' and 'type' of delivery as new memory; in addition, we should not confuse 'volume' with new memory. When new media delivers old messages quickly, in vast volumes, what we are receiving is the latest, quickest, biggest 'old' memory. New technology is a more powerful disseminator of old memory, as opposed to being some sort of automatic repository of 'new' memory. Which is not to say it can't be utilised for disseminating new memory. Equally, 'old' technology can be used for disseminating new memory. Understanding the difference between new and old memory, in relation to new and old technology, helps us avoid the trap of technological determinism and, hence, further ecological degradation.

From Stone Age to Digital Age the technological evolution of the synthetic cortex predates each newborn mind. Each new mind learns to remember the ways of the synthetic cortex of the day, prior to the cognitive recognition that she or he is imbibing other people's memory (family, community, coloniser, king, shareholder or CEO). As we have seen, as children interact with television,

Internet, smartphones, billboards, etc. they do so prior to their ability to fully contextualise the system operating the synthetic cortex. For example, traditional Aboriginal Australians learnt, via old technology, that the Earth is their mother and that animals are spiritually interconnected (Mantonvani 2000); a contemporary suburban child learns, via new technology, reverence towards shopping (Schor 2004).

The building blocks of our personal memory, the DNA-like manner of our thinking, are based on the unquestioned assumptions that informed our malleable internal cortex prior to cognitive awareness of the synthetic cortex. The *reminding* programmed into the contemporary synthetic cortex presents a global narrative, one that is dominated by a multinational corporate cultural memory. The vast majority of our playing, switching, clicking, reading and unwrapping reveals corporate memory cues: a cosmology of exploitation and consumption – one that places economic profit and personal success ahead of global ecological sustainability.

We are reminded to consume: 24/7.

We are seldom reminded to live a sustainable life. Such programming was not part of the old *Buy Now Pay Later* code written in 1944. The challenge that confronts all who upload code, art and craft onto the synthetic cortex (actor, architect, author, etc.) is how to redirect memory away from the Bretton Woods definition of the Earth. To do so, the synthetic cortex has to be embossed with flourishing reminders of the Earth's centrality to our existence, from TV soap to iPhone app. If we can achieve such re-minding, then we can bring about a sustainable, global culture; one in which the biological to cultural symbiosis of an individual is ecologically grounded due to sustainable memory cues. The challenge is substantial. Understanding the power and influence of the synthetic cortex is the first step.

> Blood, urine, ochre and sap having been heated or cooled, stretched or compressed – or simply laid bare – was mixed and fixed to rock, wood and fibre. Having been scraped or brushed, chiselled or smoothed, sown or scribed, meaning became entwined with body and other. This melding, that had no precedent and has no foreseeable end, has become our reality: the cloth that awaits our birth and the shroud that lays us down.
>
> (Fell, 2010)

References

Bednarik, R. G. (1990) 'On the cognitive development of hominids', *Man and Environment*, 15(2): 1–7.

Broks, P. (2003) *Into the Silent Land: Travels in Neuropsychology*, New York: Atlantic Monthly Press.

Campbell, J. (1972) *Myths to Live By*, New York: The Viking Press.

Coward, H. and Maguire, D.C. (eds) (2000) *Visions of a New Earth: Religious Perspectives on Population, Consumption, and Ecology*, Albany, NY: State University of New York Press.

Dissanayake, E. (1992) *Homo Aestheticus: Where Art Comes From and Why*, Seattle, WA: University of Washington Press.

Fell, B. (2005) *The Question Concerning the World Problematique and Commercial Television*, CAES Conference Scholarship & Community, ISBN 1 74108 127 0 (reprinted at www.brucefell.com).

Fell, B. (2009) *Television & Climate Change: The Season Finale,* Germany: Verlag.

Fell, B. (2010) *The Synthetic Cortex*, www.brucefell.com, accessed 17 November 2010.

Greenfield, S. (2008) *The Quest for Identity in the 21st Century*, London: Sceptre.

Heidegger, M. (1977) *The Question Concerning Technology and Other Essays*, New York: Harper & Row.

Hodgson, D. (2000) 'Art, perception and information processing: An evolutionary perspective', *Rock Art Research*, May, 17(1): 3–34.

Lehrer, J. (2009) The iPhone Mind, *The Frontal Cortex*, http://scienceblogs.com/cortex/2009/01/the_iphone_mind.php, accessed 20 March 2010.

Lipsitz, G. (1990) *Time Passages: Collective Memory and American Popular Culture*, Minneapolis, MN: University of Minnesota Press.

Lynch, G. and Granger, R. (2008) *Big Brain: The Origins and Future of Human Intelligence*, New York: Palgrave Macmillan.

Mantonvani, G. (2000) *Exploring Borders: Understanding Culture and Psychology*, Philadelphia, PA: Routledge.

Mitchell, J. (1970) 'Big Yellow Taxi', Columbia Records.

Schor, J. (2004) *Born to Buy: The Commercialized Child and the New Consumer Culture*, New York: Scribner.

Sternberg, E. (2009) *Healing Spaces: The Science of Place and Well-being*, Cambridge, MA: Harvard University Press.

13

STORY-MAKING AND MYTH-MAKING – THE PLACE OF POETIC UNDERSTANDING WHEN WRESTLING WITH REAL-WORLD PROBLEMS

David Russell

The history of Western philosophy can be understood as a dynamic tension between seeing a literal world and a literal self as agent in this world, and an as-if world and an as-if self. The notion of 'as if' is an old one in philosophy and is particularly important when we wish to describe how we feel about our relationship to both our multifaceted milieu, the one that we have co-evolved with, and ourselves as actors in the world.

It is my contention that under conditions of stress we readily fall into a simplistic way of thinking by concretising and objectifying all that we have dealings with, including how we see ourselves. To be silent in these times of ecological crisis; to be silent about the conscious dispossession of indigenous lands in Australia; to be silent about the deprivation of education from women in many parts of the world … all of these great silences exhibit a literal attitude to cultural traditions of thinking and practice. Perhaps it is in times of stress and threat that we, evolutionally speaking, have valued our ability to recognise simple distinctions and act decisively believing that the end justifies the means, as if we could see with clairvoyant eyes just what the end was and that acting with a certainty of belief was what was required by the critical nature of the events at hand.

Decisive action and certainty of beliefs look attractive in times of crisis and leaders who demonstrate the tough metal of their convictions look like strong leaders worth following. So let's assume that we have these abilities to lead and follow in times of pressing need and that these skills have proved to be evolutionally beneficial. We also have extraordinary abilities to dream into events, past, present and into the future; to see through the surface of things to the dynamic unity of all life; to touch and be touched by each and every encounter; to appreciate the poetic in the apparent mundane; and, finally, to imbue the world with extraordinary passion. It would appear that our cognitive abilities are far greater, as measured by our neuronal capacity, to perform these activities than those more akin to literal decision making.

Each generation shapes its collective imagining into a narrative that we inhabit; a story of meaning-making that both sustains and confronts us. Contemporary society offers a precarious image of stability, sustainability, responsibility: we manage and organise our worlds not only by building cities and doing science but by telling stories that serve to control, expand and explain our daily living.

The event of climate change is the one that is currently occupying our minds and the newness of such a challenge is throwing up much more of the literal than the imaginative when it comes to describing what is and what might be. What is it that we are to do? We have a taken-for-granted attitude to what we have in Australia. In fact we are used to 'the Great Australian Silence'[1] that has so successfully neutralised the crimes committed by the white settlers; the shedding of innocent blood, the denigration of Aboriginal culture, and the near total dispossession of peoples from their traditional lands. Given that we have worked hard not to feel the consequences of achieving our 'unblemished history' it is not a good omen for how we will engage with this climate change event.

An aspect of our Australian mythology is that we have no need to be reflective and certainly not to reflect on the inherent ambivalence of our emotional drivers: it is as though we are a people that love to talk but are suspect of language. Ambivalence creates uncertainty and, as the popular song of the 1980s expressed, *how much can a Koala bear*? Yet it is the emotional life that is the source of our imagination, that gives direction to our desires and, from beginning to end, gives us a reason to live and, when necessary, keep living under the most difficult of circumstances. And so it is that I am arguing for a systematic engagement with our emotional life, the images that we create that are awash with emotion, the emotionally rich stories that we tell ourselves and that constitute the very basis of our identity. If we, inevitably, make ourselves up then let's do so with a conscious passion (despite the unavoidable ambivalence of passion).

Ambivalence is a testing experience and a part of our consciousness wants to avoid it and not be disturbed by its implications. Moving quickly into our action mode in circumstances of uncertainty is a tried and true strategy of avoidance: what do I need to do, just tell me! Or, we respond to the need to be highly rational and move towards strategic planning: there must be a way out of this dilemma and, together, we can work out the best of possible action plans. Sounds appropriate but what is the path not taken, not explored. Psychologically speaking, in order to get an image that moves us, that disturbs our status quo, we need to look into those places where the majority are not looking, not seeking an explanation, and thus are actively avoiding. To live a psychological life, and I'm assuming that world circumstances are calling us to do just this right now, then one shapes, makes up, one's identity as a narrative mix of imagination and living and acting under conditions of disturbing uncertainty. Nothing less will offer us what we need for this is the soul food that has always been sought after, so mythological understanding across diverse cultures consistently reminds us, in times of deep stress. Especially so when all that is obvious, i.e. literal, has been tried and found wanting.

The most literal or concrete concept that preoccupies our mind in the contemporary world is that of 'self'; self as me, the embodiment of my experience; self understanding; self worth; self image and so on. It is not an exaggeration to say that we are obsessed with this concept of who we are and the corollary, the importance or positive regard of 'me'. The myth that we live day in and day out has my self-worth at its centre. We have made this mythic narrative stripped bare of imaginative processes. It's me, it's what's happening to me, my future, my children's future, my children's children's future. The myth of a self of significant importance is a socially constructed self that seduces the mind towards the broader myth of stability and desired certainty. As children of the Enlightenment we have created self as the object of a drive and as a positive identity to fill the experiential void inside us. The objectification and commodification of self are paralleled in our attitude to the world. The experience of no-thing-ness is simply too much to bear.

Yet, to experience a psychological life we know that we need to pursue the work of the imagination and we know that this work begins with a disturbance of the status quo. Carl Jung, the psychologist of the imagination, reminds us that it is through our woundedness, i.e. the woundedness of the ego, that we begin to see with the eyes of the soul (rather than only through the eyes of sense perception). That we begin to see into the soul of the world, *anima mundi*, the early Western philosophers called it. When we begin to doubt the certainty of the self and accept our hopelessly compromised egoic, and heroic, nature then, quite paradoxically, we engage our imaginative resources that become available to dynamically braid with our rational and energetic resources to enliven our identities and our worlds of experience. I use the plural nouns on purpose in order to reflect the plurality of styles of consciousness. Greek mythology, and all deep cultural stories, with their dynamic array of gods and goddesses and their various progeny (Apollo, Hermes, Dionysus, Artemis, Athene etc.) encouraged existential possibilities … diverse ways of attending, anything but a single consciousness, a single and unified self, one fully responsible for the consequences of all our actions and non-actions.

James Hillman, Jungian author and consummate disturber of the taken-for-granted psychological world, encourages us to gain a 'sense of the world as an animated being, as a living animal … The animal sense engages the world as highly specific and particular. This William James called 'eachness' rather than wholeness and generalities' (Hillman 2007: 151).

As an imaginative strategy, Hillman encourages us to begin by

> trusting the heart's reactions of desire and anger … These deep emotions of the bowels, liver, genital, and heart, these responses of the animal blood keep us in tune, in touch with the world around – its beauty, its insult, its danger'.
> (Hillman, 2007: 151)

There must be many ways of engaging our imagination in the task of wrestling with real world problems. And, if we are going to do this psychological work, then we need to continuously remind ourselves that the unconscious processes, those avenues of the creative disturbance, are always located where we are not looking, that is, not looking with our eyes shaped to perecive only the accepted reality.

A good starting point would be to stop 'imagining the earth as a good mother, passive, nurturing, and supportive, and to recognize the *idea* of earth to be a complex phenomenon requiring efforts of thought and imagination' (Hillman 2007: 319). This is what we humans do well when we need to: bring our imagination into active relationship with our thoughts. Our experience is then fashioned by the emotional driver that we cal *desire*, the Eros of early Greek literature. The experience of Eros is a dramatic ambivalent emotion … a bittersweet experience … desire moves, *ex-movere*, things change in the day-to-day world and we have reason to hope.

Note

1 A phrase use by the anthropologist W.E.H. Stanner to describe the deliberate attitude of all Australians in their desire to understand their history unblemished by the huge loss of innocent blood incurred in the frontier violence as the white settlers moved further and further into the indigenous lands (see Robert Manne's comment, Manne 2009).

References

Hillman, J. (2007) *Mythic Figures: Uniform Edition of the Writings of James Hillman*, Volume 6.1, Putnam, CT: Spring.

Mann, R. (2009) 'The history wars', *The Monthly*, November: 15–17.

Tayshus, A. (1983) 'Australiana', Little Digger Records.

PART 3

EDUCATION AND TRANSFORMATION

14

HENRY THOREAU

Holistic thinker, environmental educator

John P. Miller

It is only when we forget all our learning that we begin to know.

In antebellum America, a group of individuals living mostly in Concord, Massachusetts, have had a continuing influence on our lives. Daniel Walker Howe (2009) in his Pulitzer Prize winning history of the antebellum period in America, *What Hath God Wrought*, writes about the 'extraordinary outburst of genius' that was comparable in his view to fifth-century Athens or sixteenth-century Florence. He also writes of their continuing impact:

> The writings of the Transcendentalists affirm some of the best qualities characteristic of American civilization: self-reliance, willingness to question authority, a quest for spiritual nourishment. Their writings, even today, urge us to independent reflection in the face of fads, conformity, blind partisanship, and mindless consumerism.
>
> (2009: 626)

This group included Ralph Waldo Emerson, Margaret Fuller, Bronson Alcott, Nathaniel Hawthorne and Henry David Thoreau. Thoreau has inspired environmentalists for more than a century and a half as his *Walden* along with other essays continued to be cited as we face environmental degradation around the globe. Lewis Hyde in his introduction to a collection of Thoreau essays states that, 'Thoreau is arguably the original American ecologist' (2002: xxxvii). However, Thoreau was not only an environmentalist but an educator. In this paper I explore Thoreau's view of learning and education and how he was one of the first environmental educators. Thoreau loved learning throughout his life. Richardson in his book on Thoreau writes about this quality near the end of his life:

> Thoreau's nearly limitless capacity for being interested is one of the most unusual and attractive things about him. That his interests were still expanding, his wonders still green, his capacity for observation, expression

> and connection still growing is the most impressive evidence possible that his spirits this January were still on the wing.
>
> (1986: 376)

His continued interest in observation, knowing and learning can contribute enormously to our understanding of environmental education and holistic education.

Gruenewald, an educator, has argued that:

> I believe that the troubled profession of teaching could benefit greatly from taking seriously the kind of dissent, experimentation, and holistic living-in-place that Thoreau's legacy embodies. We may need him today more than ever before. So let us consider the ways in which we spend our lives, and let that reflection shape the kind of education we make possible for ourselves, and our students.
>
> (2002: 539)

In this chapter I explore how we might shape education by reflecting on Thoreau's work as a teacher and environmental educator.

His youth and teaching career

Thoreau was born and died in Concord. Although he did travel to places such as Maine and Minnesota, Concord was the centre of his life. He never tired of walking in the woods and fields surrounding the town and observing nature. He attended Harvard College and upon graduation sought a job as a teacher. He did not meet with much success in his job search so he opened his own school in Concord. In February he was able to enlist his brother, John to teach in the new school, which was called The Concord Academy. This school was a success as enrolment reached twenty-five and eventually included Louisa May Alcott, author of *Little Women*, as one its students.

The school day began with prayers and then a short talk by one of the brothers. One talk was on the change of the seasons and it was reported that the students were transported by Henry's storytelling. There were two rooms and John taught downstairs where he instructed the students in English and mathematics. Upstairs Henry taught Latin, Greek, French, physics, natural philosophy and natural history.

Harding in his biography of Thoreau states that the school was 'noted for its innovations' (1982: 82). It employed the principle of learning by doing as the students went on field trips around Concord. Henry was able to demonstrate his knowledge of the flowers and animals. Once he picked up a plant that was miniscule and, with a magnifying glass, showed the students a tiny blossom. Through these trips he was able to show his delight in nature thus hoping to inspire in his students a similar joy. The field trips were also opportunities to discover the natural history of the area. Sanborn describes one of these activities in his early biography of Thoreau.

> Henry Thoreau called attention to a spot on the rivershore, where he fancied the Indians had made their fires, and perhaps had a fishing village … 'Do you see,' said Henry, 'anything here that would be likely to attract Indians to this spot?' One boy said, 'Why, here is the river for their fishing'; another pointed to the woodland near by, which could give them game. 'Well, is there anything else?' pointing out a small rivulet that must come, he said, from a spring not far off, which could furnish water cooler than the river in the summer; and a hillside above it that would keep off the north and northwest wind in winter. Then, moving inland a little farther, and looking carefully about, he struck his spade several times, without result. Presently, when the boys began to think their young teacher and guide was mistaken, his spade struck a stone. Moving forward a foot or two, he set his spade in again, stuck another stone, and began to dig in a circle. He soon uncovered the red, fire-marked stones of the long-disused Indian fireplace; thus proving that he had been right in his conjecture. Having settled the point, he carefully covered up his find and replaced the turf, — not wishing to have the domestic altar of the aborigines profaned by mere curiosity.
>
> (1917: 205–6)

Other activities included going to the local newspaper and watching the setting of the type. The brothers also had the land plowed so that the students could plant crops in their own individual plots. Henry Thoreau also had the students do some surveying as a way to understand the practical application of mathematics. The brothers extended the recess from the usual ten minutes to thirty minutes to allow the students more freedom. Another example of how relaxed the brothers were with the students is when Henry and John put tar on their boat the students watched them and played in the shallow water (Harding, 1982: 83).

The school ran for three years and was closed on 1 April 1841, when John's health started to fail. Harding writes that 'The pupils remembered the schools and its teachers with "affection", "gratitude", and "enthusiasm"'. One of the students, Benjamin Lee, thought he would never forget the kindness and goodwill of the Thoreau brothers 'in their great desire to impress upon the minds of their scholars to do right always' (Harding, 1982: 86). Harding goes on to assert that:

> The Thoreau school was a century ahead of its time. Granted Bronson Alcott's famous Temple School had anticipated some of its innovations by a few years, but Alcott with his experimentation had brought the wrath of the community down on his head. The Thoreau's, on the other hand, although many of their innovations were more radical than those of Alcott, won acceptance and made of their school a considerable and memorable success.
>
> (1982: 86)

John died in early 1842. The death of his brother and closest friend was devastating for Thoreau. Yet Richardson argues that after a period of illness and mourning Thoreau gained a sense of freedom that led to creative output in his writing. The death of Emerson's young boy, Waldo, who was close to Thoreau, also seemed to spark a sense of urgency and self-discovery.

Walden

Thoreau's two-year stay at Walden Pond and the book that resulted from that period were defining marks in Thoreau's life. Richardson writes that Thoreau produced the greatest amount of quality writing during his stay there. Emerson owned the land on which Thoreau built his small cabin. Thoreau went to Walden for several reasons. First, he wanted independence as, outside of his years at Harvard, he had always lived with either his family or at Emerson's. Second it was a statement of social reform. This was a time when communal living was popular with projects such as Brook Farm and Fruitlands; in contrast, Thoreau wanted to explore a smaller, simpler form of living (Richardson 1986).

At Walden he was able to practice his own form of contemplation as he sat in front of his cottage.

> I sat in my sunny doorway from sunrise till noon, rapt in a revery, amidst the pines and hickories and sumachs, in undisturbed solitude and stillness, while the birds sang around or flitted noiseless through the house, until by the sun falling in at my west window, or the noise of some traveller's wagon on the highway, I was reminded of the lapse of time. I grew in those seasons like corn in the night, and they were far better than any work of the hands would have been. They were not time subtracted from my life, but so much over and above my usual allowance. I realised what the Orientals mean by contemplation and the forsaking of works. For the most part, I minded not how the hours went.
>
> (Thoreau, 1986: 157)

Contemplation was a way of knowing what was important to Thoreau. In his journal he described it in more detail:

> I must walk more with free senses. It is as bad to *study* stars and clouds as flowers and stones. I must let my senses wander as my thoughts, my eyes see without looking … Be not preoccupied with looking. Go not to the object; let it come to you. When I have found myself ever looking down and confining my gaze to the flowers, I have thought it might be well to get into the habit of observing the clouds as a corrective; but no! that study would be just as bad. What I need is not to look at all, but a true sauntering of the eye.
>
> (Shepard, 1961: 99)

Contemplation, in Thoreau's view, allows the person to *be* with an object rather than becoming a subject observing an object. This way of knowing is generally ignored in education but needs to be included if we are truly to have a more holistic approach to learning.

Walden is meant to inspire the individual to awaken and to live the life they can imagine. Thoreau encourages us to 'hear a different drummer'. Walden is read in many high school classrooms but one wonders whether the educational system with its emphasis on standardisation actually undermines this inspiring vision.

The transcendental naturalist

As Thoreau matured into his mid- to later thirties he took more and more interest in science and the observation of nature. His later journals are filled with his observation of nature that he gathered on his walks around Concord. He wrote in his journal that he found himself 'inspecting little granules, as it were, on the bark of trees ... and I call it studying lichens' (Shepard 1961: 81). These walks were precious to him and he wrote in his journal that he much preferred to walk alone. He wrote,

> They do not consider that the wood-path and the boat are my studio, where I maintain a sacred solitude and cannot admit promiscuous company ... Ask me for a certain number of dollars if you will, but do not ask me for my afternoons.
>
> (Harding, 1982: 292)

Thoreau read widely throughout his life and was influenced by many naturalists and scientists including Goethe, Darwin, Agassiz and Linnaeus. Richardson notes that Linneaus wrote about the creative forces of nature and that the 'whole earth is alive, not inert' (1986: 255). Echoing this idea, Thoreau wrote in *Walden*:

> The earth is not a mere fragment of dead history, stratum upon stratum like the leaves of a book, to be studied by geologists and antiquaries chiefly, but living poetry like the leaves of a tree, which precede flowers and fruit, not a fossil earth, but a living earth; compared with whose central life all animal and vegetable life is merely parasitic.
>
> (1986: 357)

There are hints here of the Gaia hypothesis that suggests that entire earth is a living organism with its own intelligence. In his journal Thoreau wrote that science 'sees everywhere the traces, and it is itself the agent, of a Universal Intelligence' (Shepard 1961: 125). Although in his later years Thoreau focused on detail, it was always against the background that nature operated according to universal principles. The human mind also reflects these universal principles. In writing about husbandry in his journal from the ancients to today he writes about the 'perennial mind' that does not change from one year to another. The 'perennial mind' is that deep wisdom within the individual, which is in harmony with nature.

In 1855 Frank Sanborn opened a new school in Concord. He asked Thoreau to lecture at his school once a week, but Thoreau was concerned that it would take away from his other tasks. Sanborn took the students on weekly excursions around Concord and occasionally Thoreau would join them. In one of his books on Thoreau, Sanborn cites one student, Samuel Higginson, who recalled Thoreau. Higginson wrote:

> He was … to us more than a charming companion; he became our instructor, full of wisdom and consideration, patiently listening to our crude ideas of Nature's laws and to our juvenile philosophy, not without a smile, yet in a moment ready to correct and set us right again. And so in the afternoon walk, or the long holiday jaunt, he opened to our unconscious eyes a thousand beauties of the earth and air, and taught us to admire and appreciate all that was impressive and beautiful in the natural world around us. When with him, objects before so tame acquired new life and interest. We saw no beauty in the note of veery or wood-thrush until he pointed out to us their sad yet fascinating melancholy … He turned our hearts toward every flower, revealing to us the haunts of rhodora and arethusa.
>
> (1862: 313)

Synthesis

Recent research on Thoreau has found that in his last years that he was integrating his inquiries in nature with science, literature and philosophy to form a 'comprehensive theory of the process of nature's variation and development … the pursuit of the knowledge of a single thing became the unceasing quest to comprehend the encompassing unity of all things'(Robinson, 2004: 177–8). Robinson argues that Thoreau's work focused on an:

> ever-enlarging network of relations, which natural objects were defined through their part in a larger system, and thus through the process of their interactions. In accord wit his long-held faith that the study of nature was intimately connected with the culture of the self, Thoreau believed that this ever-enlarging system of relations also included human consciousness and human agency
>
> (2004: 184)

Thoreau was seeking what he called 'Beautiful Knowledge'. Thoreau wrote: 'The highest that we can attain to is not Knowledge, but Sympathy with Intelligence' (Hyde 2002: 23). His metaphor is the 'lighting up of the mist by the sun'. Thoreau used the sun as an image of that Intelligence that is the source of Beautiful Knowledge. At the end of the essay 'Walking' he writes:

> So we saunter toward the Holy Land, till one day the sun shall shine more brightly than ever he has done, shall perchance shine into our minds and ears,

> and light up our whole lives with a great awakening light, as warm and serene and golden as on a bankside in autumn.
>
> (Hyde, 2002: 28)

Although the closure of the Concord Academy was the end of Thoreau's teaching career, he wrote about education and learning throughout his life. Some of the entries in his journal toward the end of his life are about education. For example, he wrote this in 1859, about two and half years before he died, 'How vain to try to teach youth, or anybody truths! They can only learn them after their own fashion, and when they get ready' (Shepard 1961: 212). Thoreau was anticipating educators like A. S. Neill and Carl Rogers with such thoughts.

Thoreau's contribution to education

Environmental education

Thoreau can be seen as the first environmental educator. First, there was the work he did in Concord Academy with the field trips into nature. He continued this work in Sanborn's school as well.

Walden and his other writings have inspired environmentalists and conservationists for over 150 years. One of today's leading environmental educators, David Orr writes this about Thoreau.

> *Walden* is a model of the possible unity between personhood, pedagogy and place. For Thoreau, Walden was more than his location. It was a laboratory for observation and experimentation; a library of data about geology, history, flora and fauna; a source of inspiration and renewal; and a testing ground for the man. *Walden* is no monologue; it is a dialogue between a man and place. In sense, *Walden* wrote Thoreau. His genius, I think, was to allow himself to be shaped by his place, to allow it to speak with his voice.
>
> (1992: 126)

Unlearning

Thoreau was an advocate of letting go of our previous learning so that we can see clearly. One of the early entries in his journal describes this process:

> As the least drop of wine tinges the whole goblet, so the least particle of truth colours our whole life. It is never isolated, or simply added as treasure to our stock. When any real progress is made, we unlearn and learn anew what we thought we knew before.
>
> (Shepard, 1961: 3)

He continued this theme throughout his life and towards the end of life he wrote this in his journal:

> It is only when we forget all our learning that we begin to know. I do not get nearer by a hair's breadth to any natural object so long as I presume that I have an introduction to it from some learned man. To conceive of it with a total apprehension I must for the thousandth time approach it as something totally strange. If you would make acquaintance with the ferns you must forget your botany.
>
> (Shepard, 1961: 210)

One is reminded of Eastern approaches to learning: for example, Suzuki said that we need a *beginner's mind* to see things as they really are (1970). Learning then is not filling up the mind with facts and concepts but being able to see clearly. Clarity of vision then can lead to wisdom and a deeper understanding of the environment.

Learning by doing

I have already cited Harding's comment on Thoreau's pedagogy, which focused on learning by doing rather than listening to lectures. In *A Week on the Concord and Merrimack Rivers* Thoreau wrote, 'Knowledge is to be acquired only by a corresponding experience. How can we *know* what we are *told* merely?' (1980: 365).

Thoreau gives an example of learning by doing in *Walden.* He asserts that students should not just study life but *live* it from beginning to end. 'Which would have advanced the most at the end of a month, — the boy who had made his own jackknife from the ore which he had dug and smelted, reading as much as would be necessary for this, — or the boy who attended the lectures on metallurgy?' (1986: 94–5).

Embodied knowing

In his eulogy on Thoreau, Emerson commented on how important physical activity was in his life, 'there was a wonderful fitness of body and mind … The length of his walk uniformly made the length of his writing. If shut up in the house, he did not write at all' (2003: 460). Thoreau himself wrote, 'A man thinks as well though his legs and arms as his brain. We exaggerate the importance and exclusiveness of the headquarters' (Shepard 1961: 212). Environmental education cannot be just a head trip, it must be rooted in our bodies as well.

The senses are the way into what Thoreau called the natural life. He wrote, 'We pray for no higher heaven than the pure senses can furnish, purely sensuous life' (1980: 379). For Thoreau the senses were a way to connect to the divine.

> The ears are made, not for such trivial uses as men are wont to suppose, but to hear celestial sounds. The eyes were not made for such groveling put to and worn out by, but to behold beauty now invisible. May we not see God.
>
> (1980: 382)

Thoreau then called for an education of the senses. 'What is it, then to educate but develop these divine germs called the senses?' (1980: 382).

However, Thoreau was not a materialist. He encouraged inquiry into what he called the 'OTHER WORLD which the instinct of mankind has so long predicted'. He felt that 'Menu, Zoroaster, Socrates, Christ, Shakespeare, Swedenborg' were the 'astronomers' of that world that could guide in that inquiry (1980: 386).

Community as the school

Thoreau believed that learning should not be limited to the school building but that the entire community should participate in the children's education. In *Walden* he makes the call for 'uncommon schools' so that 'villages were universities' (1986: 154). He also believed that learning should continue throughout our adult lives.

Relationship with students

When Thoreau was searching for a teaching position, he wrote to Orestes Brownson about his vision of teaching.

> I would make education a pleasant thing both to the teacher and scholar. This discipline, which we allow to be the end of life, should not be one thing in the schoolroom and another in the street. We should seek to be fellow students with the pupil, and we should learn of as well as with him, if we would be most helpful to him.
>
> (Harding, 1982: 55)

The comments from former pupils, cited earlier, would indicate that Thoreau treated his students with respect and care. Although he could be difficult in his relationships with adults, he enjoyed being with children. Throughout his life Thoreau loved demonstrating nature's beauty and wonder to children. When Thoreau died, school in Concord was dismissed and many children attended the funeral.

We can be thankful that Thoreau let himself stand in the light. Thoreau not only developed ideas about non-violent resistance that inspired Gandhi and King, held a deep respect for the earth and its processes, but he also created a holistic pedagogy that provides a foundation for environmental education. If we took Thoreau's work seriously as educators and brought his ideas into teaching, it would do a great deal to move away from education systems that are obsessed with individual achievement and testing. More importantly, it could help us move forward in addressing our massive environmental problems.

I close this paper with a quotation from a letter to Harrison Blake with whom Thoreau carried on a life long correspondence:

> Do what you love. Know your own bone; gnaw at it bury it, unearth it, and gnaw it still. Do not be too moral. You may cheat yourself out of much life so. Aim above morality. Be not *simply* good — be good for something. — All fables indeed have their morals, but the innocent enjoy the story. Let nothing come between you and the light.
>
> Thoreau, 2004: 38)

References

Emerson, R. W. (2003) *Selected Writings of Ralph Waldo Emerson*, New York: Signet.

Gruenewald, D, A (2002) 'Teaching and learning with Thoreau: Honoring critique, experimentation, wholeness and the places where we live', *Harvard Educational Review* 72(4): 515–41.

Harding, W. (1982) *The Days of Henry Thoreau: A Biography*, New York: Dover.

Higginson, S. S. (1862) 'Henry D. Thoreau', *Harvard Magazine*, VIII: 313–18.

Howe, D. W. (2007) *What Hath God Wrought: The Transformation of America, 1815–1848*, New York: Oxford University Press.

Hyde, L. (ed.) (2002) *The Essays of Henry D. Thoreau*, New York: North Point Press.

Orr, D. W. (1992) *Ecological Literacy*, Albany, NY: State University of New York Press.

Richardson, R. D. (1986) *Henry David Thoreau: A Life of the Mind*, Berkeley, CA: University of California Press.

Robinson, D. M. (2004) *Natural life: Thoreau's Worldly Transcendentalism*, Ithaca, NY: Cornell University Press.

Sanborn, F. B. (1917) *The Life of Henry David Thoreau*, Boston, MA: Houghton Mifflin.

Shepard, O. (ed.) (1961) *The Heart of Thoreau's Journals*, New York: Dover.

Suzuki, S. (1970) *Zen Mind, Beginner's Mind*, New York: Weatherhill.

Thoreau, H. D. (1980) *A Week on the Concord and Merrimack Rivers*, Princeton, NJ: Princeton University Press.

Thoreau, H. D. (1986) *Walden and Civil Disobedience*, New York: Penguin.

Thoreau, H. D. (2004) *Letters to a Spiritual Seeker*, New York: W.W. Norton.

15

A CURRICULUM OF GIVING FOR STUDENT WELLBEING AND ACHIEVEMENT – *'HOW TO WEAR LEATHER SANDALS ON A ROUGH SURFACE'*

Thomas William Nielsen

The whole world is not nice!

Sitting in my office, reading the feedback that my pre-service teaching students hand in at the end of every semester, I was given a challenge. A student praised the way that I conducted classes and my positive attitude towards students and life in general but ended the comments with the words: '... *but the whole world is not nice, man!*'

The challenge that this student gave me was to reconcile the fact that one can be positive and well meaning but that that doesn't mean that the world will reciprocate. The entire semester I had been showing my students what the research tells us about the effects of positive emotions and notions of values education and community service. Now this student pointed out a possible flaw in my rationale for teaching this to beginning educators, and arguing, as I do, that perhaps we need to be aware of these dimensions in order to increase the general wellbeing of students. If the whole world is not nice, perhaps we can indeed set children up for much disappointment and even assist in the development of a type of 'naive happiness' that some are warning against (Forgas 2007; Schnall et al. 2008; Hamzelou 2010).

The state of childhood: A reflection of us all

In positing the view that it is beneficial to teach children to think positively and act generously even when this doesn't guarantee reciprocity from their surroundings, I will provide some background information about the larger context. There is now sufficient and, in my view, incontrovertible evidence-based research on the topic to be concerned with the health and happiness of young people (see especially Stanley in Childs et al. 2008; World Health Organization 2008). Obesity and allergies have been on the rise for some time. Mental health problems also affect a lot more now. Depression and suicide rates have risen steadily in the past

60 years. Violence, antisocial behavior and binge drinking among young people are now so prevalent as to be viewed by some as the norm rather than the exception. Moreover, all of the above phenomena are now being observed in younger age brackets across the full spectra of socio-economic strata and demographics.

The innocence of children is also becoming increasingly harder to protect in our highly visual and readily accessible virtual world. Violent and sexually explicit images have been shown to have a disturbing effect on the development of children (Braun-Courville and Rojas 2009), while Internet filters in schools and in homes remain an imperfect Band-Aid solution at best. No doubt comparable to most Western countries, by the time they are eleven, American children will have seen 8,000 murders and 100,000 acts of violence on TV (Huston et al. 1992), while nearly 3,500 studies worldwide show a clear correlation between violent behavior and exposure to screen violence (Grossman and DeGaetano 1999).

The obvious challenges facing the present state of childhood (which I take to include the teenage years) in turn have ramifications for educators and parents and, in many ways, our society in general. The teacher attrition rate, particularly among entry-level teachers, is high in most Western countries (Ewing and Smith 2003), with many teachers and researchers citing behavior problems and student discipline as the main reason for leaving the occupation. Problem behavior is also a major cause for concern in parents and a common reason for family unrest. While problem behavior in children can have many triggers, a common denominator seems to be underlying states of unhappiness and lack of wellbeing (Dolan et al. 2008).

What does all this mean for educators, parents and society in general? What can be done to turn the statistics around? Surgery for obesity seems extreme, and curative therapy against depression is uncertain at best. Bans against alcohol and other unhealthy practices often seem to make those activities even more interesting to young people. Working with teachers and parents on a regular basis, I can also testify to the fact that there is no silver bullet or universal panacea with which to cure stressed carers of children either. We seem to have few answers to our current problems – at least not if we only look toward remedial strategies.

Values education, quality teaching and service learning

Because curative practices are uncertain at best, more and more educational settings are looking to preventive measures for increasing children's wellbeing and resilience. Notions of values education, character education, resilience education, positive education, civics education, social emotional learning, etc. are being implemented in many countries at the moment. While these initiatives each have their own emphases and research traditions, they all share the hope that through preventive measures we can increase children's general wellbeing, resilience and better judgment.

A comprehensive meta-analysis of the literature (see Lovat et al. 2009) reveals that notions of values education are predictive of increased individual and communal wellbeing, pro-social behavior and classroom ambience (Benninga et al. 2003, 2006; Berkowitz and Bier 2004, 2007; Deakin Crick et al. 2004, 2005; Billig 2007). We now also have a body of evidence that links pro-social behavior with academic success (Wentzel 1991a, 1991b, 1993, 1996; Juvonen and Wentzel 1996; Caprara et al. 2000; Welsh et al. 2001). In other words, when combined with the tested and tried tools of quality teaching (Lovat and Toomey 2007), which ensures that educators do not revert to outdated models of transmission teaching or moral inculcation, notions of values education and social emotional learning become intrinsically linked to wellbeing, pro-social behavior and academic success. As a result, this type of education should no longer be seen as a moral imperative but rather a pedagogical imperative (Lovat et al. 2009); that is, quality teaching that explicitly addresses human and social values creates not only happier and more pro-social learners but also *better* learners.

There are those who argue that efforts to increase children's happiness assume that happiness is always a good thing, and it is true that there are studies that show that happy children can be less objective and overly preoccupied with self-gratification (Forgas 2007; Schnall et al. 2008; Hamzelou 2010). To fully understand this phenomenon, however, the work of Martin Seligman is of importance in highlighting that happiness is not just about self-gratification inspired happiness (Seligman and Csikszentmihalyi, 2000; Seligman, 2002a, 2002b, 2008). Seligman identifies three types of happy living: the 'pleasurable' life, when we gratify our senses (e.g. enjoying an ice cream); the 'engaged' life, when we engage our signature strengths in activities that makes us lose track of time (e.g. sport); and the 'meaningful' life, which is when we are something for others (e.g. peer tutoring). The important point to note here is that when we are happy from having meaning in our lives, we are often still using our signature strengths but for a higher purpose than our own gratification. While sensory pleasure and engaging activities are not to be dismissed, we enjoy higher and steadier levels of happiness and recuperate more easily from trauma when our lives also contain meaning.

The educational significance of this is twofold. Firstly, it is clear that the studies that have shown negative effects of being happy have focused on the type of happiness that primarily comes from pleasure, in which case it also makes sense that we can become more selfish if exposed to pleasure, because it actually focuses us on our own senses (in many ways this is a challenge facing our modern way of living in general). Even happiness deriving from engagement can, on its own, house an inherent risk of making us more insensitive to others (e.g. a teenager who is engaged in a computer game for hours). However, with happiness that stems from having meaning in our lives, which is defined as deriving from being something to others, we notice that this level of wellbeing by definition *connects* us to others.

Realizing the qualitative differences between states of happiness in turn highlights the second significance: that notions of values education and social emotional learning are potentially more transformative when they incorporate not only explicit teaching and learning about individual values, emotions and self-awareness but also activities involving service or giving to others – the meaningful life. This is a particularly effective form of values education, often referred to as 'service learning', which is when children serve others while learning the curriculum. For example, while studying the effects of drought in Australia (learning), a class might raise funds for relief packages to drought-stricken farmers (service). Again, the literature is very clear about the effects of this type of values education: service learning is predictive of increased individual and communal wellbeing, as well as academic improvement (Conrad and Hedin 1991; Billig 2000; Lovat et al. 2009). In fact, service learning is one of the types of values education that seems to have the biggest positive impact on, especially, marginalized students (Lovat et al. 2009).

Towards a curriculum of giving

Understanding that giving to others can connect the different levels of happiness with the most important level of meaning – being something for others – we also begin to understand that 'happiness', per se, is not really what we are after in schools. What we are after is a type of wellbeing, as defined by having a meaningful life, and to which many students are seldom exposed. Since the Second World War, spending power has tripled in most Western countries, while levels of happiness and psychological wellbeing have remained flat and even reduced, according to some sources (see especially Seligman 2002a). Many teenagers, when asked, do not want to give, probably because many of them have never experienced the *opportunity* to give (Townsend 1992).

Yet it is noteworthy that even when the giving is 'enforced', as in the case of service learning, the benefits remain: students who are exposed to service learning end up volunteering more afterwards (Post and Neimark 2007). In schools, giving, or service, is also a strong predictor of increased mental and physical health into adulthood and reduced adolescent depression and suicide risk (Dillon et al. 2003). Other studies confirm that giving has a significant impact on teenagers' mental health, increasing their happiness, hopefulness and social effectiveness (Billig 2000, 2007; Csikszentmihalyi 2002; Scales et al. 2006). One study (Oman and Thoresen 2000) even found that giving reduces mortality significantly; it followed almost 2,000 individuals over the age of fifty-five for five years, and those who volunteered for two or more organizations had a 44 per cent lower likelihood of dying – 14 per cent lower than those who exercised four times a week.

In a word, the research is quite clear: one of the healthiest things we can do is to give, as this leads us to being healthier, happier and possibly even living longer. This is why I prefer talking about the need for a theory and a curriculum of giving, as opposed to simply service learning, because I think the former better denotes the intrinsic value of giving and service, and that such practices will be of benefit *in and of themselves*, whether or not they are tied to the 'normal' curriculum. By considering how we as teachers can allow regular opportunities for children to have meaning in their lives via a curriculum of giving – and by being familiar with the research on giving – we are more likely to value the underlying benefit of giving, and not merely see it as a means to an end. Giving, as a principle of living, can be embodied in almost any situation and of itself has immense value to individual and collective wellbeing.

Having been an advisor and researcher in several of the Australian Government's values education projects (2004–2010), I have also personally experienced the benefits of giving, particularly as they relate to children and teenagers. Using the qualitative research methodologies of action research and meta-analysis of efforts in more than 300 schools all over the nation, one of the key findings in the Australian Values Education Final Report (Stage 2) was the recommendation that educational settings 'develop relevant and engaging values approaches connected to local and global contexts that offer opportunities for real student agency' (AGDEST 2008). In my experience of values education in Australia, the incorporation of giving into units and lesson plans have been the most transformative way of 'connecting students to local and global contexts' and creating 'real student agency' – in other words, creating meaning in the lives of students. What follows are the beginnings of a framework for educators to implement a Curriculum of Giving.

Four dimensions of giving

I have through my research on values education in Australia (e.g. Nielsen 2005; Lovat et al. 2009) come to appreciate four dimensions in which we as educators can engage learners in giving. These four dimensions encompass a continuum from self-compassion to altruism, from micro to macro cosmos (see Table 15.1).

Studies have shown that self-compassion can have a significant positive association with self-reported measures of happiness, optimism, positive affect, wisdom, personal initiative, curiosity and exploration, agreeableness, extroversion and conscientiousness (Neff and Vonk 2009; Neff et al. no date), which is why I see giving to self as an important aspect of giving; for example, we all know that it is hard to give to others if we don't have a surplus from which to give.

As to my use of 'altruism', I am here not concerned with whether innate altruism exists or not. My use of the term is purely within the understanding that when we as humans give to others, the research indicates that it is one of the healthiest things we can do for our individual and collective wellbeing – whether

Table 15.1 Four dimensions of giving

Self-compassion (micro)		*Altruism (macro)*	
Self	*Relationships*	*Communities*	*Life*
Self-soothing	Acts of kindness	Relief aid	Gratitude
Flow	Courtesy	Petitions	Awe
Savoring	Peer tutoring	Clean-ups	Poem/prayer

or not altruism is a spiritual quality, or simply a Darwinian 'survival' mechanism, useful for social capital.

I will now provide elaboration and examples of each of these dimensions. I do so in the belief that it does not matter so much that we do not know exactly why giving works, but that it is important to know that it does work and what it may look like in educational settings.

Giving to self – 'know thyself'

Schools I have worked with have implemented notions of positive psychology to help students know and appreciate themselves more. In this type of education, children have been learning about Signature Strengths (Seligman 2002a), Flow (Csikszentmihalyi 2000) and Savoring (Bryant and Veroff, in Seligman 2002a). What this does is help students to know and appreciate their innate qualities (Signature Strengths), how they can 'get lost' in activities (Flow) and enjoy the pleasures of life with moderation and awareness (Savoring).

Some schools have also taught students how to deal with trauma and notions of 'catastrophic thoughts', which is what we all can experience at times when an incident sets off a sequence of negative thoughts that spiral our optimism and hopefulness downwards. Seligman (2002a) taught ten year olds how to dispute catastrophic thoughts and how to energize themselves, cutting their rates of depression in half as they went through adolescence. Teachers I have worked with confirm the impact of explicitly teaching how to deal with difficult thoughts:

> I've been working with [a student] for over 12 months – his father died as the result of an accident a few years ago, and he's really been struggling to overcome it … We started talking about bliss, and flow, which gave him an opportunity to talk about his memories of his dad in a pretty positive way … He really finally opened up to me and a couple of other boys … by the fire one night after everything had wound up. It was incredibly brave, and incredibly empowering. It was at this stage that I introduced the phrase from *Tuesdays With Morrie* – 'Death ends a life, not a relationship' to him – and that was a mantra that allowed him to work through some pretty tough thoughts. At this stage he stopped having recurring nightmares about his father's death – and through this 'talk', he was able to engage fully with a counselor, which made great breakthroughs for him also. It took a long time, but the turnaround in this kid has been significant …
>
> (Teacher, male, ACT)

It might be argued that teaching positive psychology for the 'self' is not really giving, but, again, I would argue that it is important that we see giving as existing on a continuum from self-compassion to altruism. Without giving to the self, with wisdom and awareness, what the self needs, it is hard to give effectively to others. Such a view also sheds light on a long-standing debate in education as to whether skills or self-esteem is more important to build up in order for the other to develop. What has been missing in this debate is the important ingredient of altruism, or a 'wellbeing psychology', because if we add this dimension we realize that one can have all the skills and self-esteem in the world and still be a bully. Self-esteem and skills, in themselves, cannot achieve a satisfactory support of the other, because they both depend on being situated in a psychology of human wellbeing – if they are to be of benefit *to* human wellbeing.

This in turn explains why bullies can have both high and low self-esteem, whereas people who stand up for victims of bullying overwhelmingly have high self-esteem. The data on self-esteem and skills, on its own, is conflicting, as people with high self-esteem and high skill sets still appear to be susceptible to becoming bullies, alcoholics, drug users and depressed. In contrast, the data on altruism is consistently linked to wellbeing and pro-social behavior (see especially Post and Neimark 2007). Indeed, self-compassion is much more predictive of positive affect and wellbeing than self-esteem (Neff and Vonk 2009; Neff et al. no date), again underlining the relevance of viewing the self as part of an altruistic continuum that stretches from the micro to the macro cosmos.

Many great teachers throughout history have argued the importance of 'knowing thy self' and that true knowledge is intimately linked with virtues (e.g. Socrates). It seems that we now have research that supports this view, which is why a curriculum of giving must start in self-awareness and self-compassion. As the above story illustrates, the language and concepts of positive psychology are particularly useful tools to this end.

Giving to relationships – from 'me' culture to 'we' culture

The more obvious way of giving for many children will be to give to their most immediate relationships, such as family, friends and peers. With this dimension, it is not so difficult to conceptualize what constitutes giving; the challenge for teachers is to allow regular opportunities for giving, since traditionally this has not been an issue that teachers have felt should concern them. As noted, however, many children do not have regular opportunities to give, and so it has become increasingly important that educational settings allow for such opportunities.

Schools I have worked with have incorporated notions of respect, courtesy and acts of kindness into classroom and schoolyard interactions. Rudolf Steiner Kindergartens, in particular, are exemplars of allowing regular and routine acts of giving for younger children. In these settings, one will often see children setting the table for morning tea and helping staff prepare it. Afterwards, some children might work in the garden, weeding and picking salad and vegetables for lunch. At lunchtime, others are again at the centre of washing, cutting and preparing the food. In the afternoon, one might see some children tidying up, folding blankets and doing other household tasks.

In older grades, some schools have explicitly focused on a particular value related to giving, such as respect or integrity, with noticeable results:

> In grade six I was, I guess, a bully. And I used to bully an overweight kid … and I only did it cos my friends were doing it … it didn't really feel right, but I did it anyway cos I didn't know what respect was, I didn't know what integrity was … Once we had a semester on … respect and learned what it was, [the bullying] pretty much just stopped … [I] stopped hanging out with the mean kids.
>
> (Year 10 student, male, ACT)

This student quote is significant because it exemplifies the power of explicit values education, providing the outer support to nourish what for many is already present inside – even if only as a quiet 'voice'. By being supported to consciously know what integrity was, this student realized that it was what he had felt inside him all the time, and that not following this inner voice made him feel 'not right'. In other words, explicit values education provided the supportive 'nudge' to become more fully in the outer psyche what he was already expressing to himself in his inner psyche.

As one senses in the above data, there are many ways to give within the dimension of children's immediate relationships, but a crucial factor is whether such learning opportunities are allowed and catered for. Whether taught explicitly or implicitly, such a curriculum could reverse trends toward the 'me' culture, as opposed to a 'we' culture, that some warn against (see, for example, Brooks 2008).

Giving to communities – expanding the 'consciousness'

It is good to expand children's consciousness even further by allowing them opportunities to give to a wider circle than just their immediate relationships. Giving to the local and global community helps transform a child's natural egocentricity toward more empathetic and global states of consciousness.

In one of the schools I worked with, two teachers, who team-taught two grade 3–4 classes, decided to let the values of care, compassion and responsibility guide a unit called 'Cool Kids 4 a Cool Climate'. One of the teachers had the students writing a diary as if they were a drought-stricken farmer. For children of this age, the results were dramatic:

> Dear Diary, Today is Friday, another busy day. The earth is all dry and very deeply cracked. The poor old sheep dog is too thirsty to round up the sheep and the cows are so skinny you can see their rib bones. I have to shoot them tomorrow. I am out of water to feed them so I think it is best … I don't know what to do. Should I sell the property of not?
>
> (Year 3–4, student, ACT)

Having developed empathic links, students were then involved in collecting relief packages for drought-stricken farmers. One of the teachers reported:

> The culminating activity ... was for the children to list ways to show that they care about the effects of drought on our community. From this list they chose, as a group, one idea to put their words into action. They chose to collect tinned food, blankets, toys and books to donate to the Country Women's Association (CWA) to go to drought affected farmers in our area. Over the space of five weeks they developed a plan, allocated jobs, and organized posters and newsletter articles. They collected over 350 tins, numerous blankets, four boxes of clothing, as well as toys and books. A representative from the CWA collected the donations and the children received a letter of appreciation. I wish I could include the photo we took of the children with their collections so you could see the look of pride on every face ...
>
> (Teacher, female, ACT)

As with the other dimensions of giving, the advantage of a hands-on approach to developing generosity, empathy and compassion is that this can happen without moralizing. Through giving to others, children can experience wonderful subtle emotions that would never come about through theoretical learning alone. In many respects, giving enables children to scale Maslow's Hierarchy of Needs and reach the self-actualization pinnacle (Maslow 1943).

Giving to life – reconnecting with the 'sacred'

A good example of giving to life is gratitude for life itself. Interestingly, similar to the health benefits deriving from overt giving, studies have shown that 5–15 minutes of gratitude will cause a shift in the nervous system to a calm state, called 'parasympathetic dominance', which is where heart, breathing, blood pressure and brain rhythm are synchronized, and where beneficial hormones increase and stress hormones decrease (see Rollin McCraty in Post and Neimark 2007).

My own research on giving is consistent with the broader research on the topic. With the consent of my pre-service teachers, in recent years, I have assigned tasks to them that have had direct impacts on their wellbeing. To validate this finding, I have randomly selected approximately a third of the 200 students that I have the pleasure of teaching each semester to do a random act of giving every day, another third to keep a daily gratitude journal, and the last third to 'just' learn about positive psychology (control group). The activities of giving and gratitude are voluntary and not assessed, but I ask all students to take a wellbeing survey (http://www.australianunitycorporate.com.au/community/auwi/Pages/default.aspx) at three points in the thirteen-week semester. Each time I have conducted this research, the gratitude and giving groups increased their wellbeing, while the control groups only maintained their original level of wellbeing (Nielsen 2010).

In other words, gratitude, as an example of 'inner giving', seems to produce similar health benefits as overt giving, and it therefore makes sense to view acts like gratitude, reverence, awe, prayer, etc. as ways of giving to life itself – no matter what one perceives 'life' to be in terms of ideology and beliefs. Schoolteachers have also told me about the noticeable effects of introducing this type of giving:

> The Year 12s of 2008 had been a truly energetic, passionate and community spirited group. Early in the year I planted the seed with the student leadership team that they could make a great impact on younger students if they were willing to show just how much school had meant to them and how thankful they were … They chose the theme of thanksgiving for their final mass. In the past, themes have been along the lines of friendship, journey, taking flight etc. Just in choosing this theme they had already changed the tone and feel of how they were to finish as a year group. In preparing for the mass they had many discussions as a team and as a whole year group about what they were grateful for and how they would like to show that gratitude … I believe that these expressions of thankfulness transformed the community … But most importantly these students had experienced the power of giving thanks … In my memory, this was the most pleasant experience in a school of a Year 12 group leaving the community and upon reflection I now see that by embracing the value of gratitude, they were able enter into the experience with authentic joy and appropriate expression.
>
> (Teacher, female, suburban NSW)

Giving to life can take many forms, illustrating that giving can be a principle of life that we, ideally, have the opportunity to engage with in almost any situation. As such, it seems to be a state of mind (and heart), which has parallels with notions of Flow (Csikszentmihalyi 2000) and Bliss (Seligman 2002a). Just as Flow and Bliss activities might be precursors to developing the kind of self-awareness and appreciation that could underpin, for example, drug education, so does giving to life – through gratitude, mindfulness, reciting poems or prayers – seem to build an awareness and a 'presence' (see Tolle 2005) that is of benefit to students in all pursuits of life.

Wearing 'leather sandals'

> You can't cover the whole world in soft leather to make your journey smooth and comfortable, but you can wear leather sandals.
>
> (Buddhist saying)

With the above elaboration on giving to life itself, I have come full circle in terms of how I would respond to the student whose feedback challenged my teaching with the statement '*... but the whole world is not nice, man!*' This is what I would now say to this student: being generous and striving for positive thoughts and emotions has immense value to the self, whether or not the world reciprocates. One cannot cover the whole world in smooth and soft leather, but one can wear leather sandals so that at least the space that one is occupying is still soft and comfortable. The other thing to note, of course, is that the more we all wear 'leather sandals', the more the world in turn will be a 'smoother' and more 'comfortable' place. Moreover, being positive and generous are not mutually exclusive to being discerning and effective; increasing our 'meaningful' wellbeing (as opposed to just 'pleasurable' happiness) seems to make us *more* capable, productive and creative (Isen et al. 1991; Seligman,= 2002a; Post and Neimark 2007).

I have in this chapter used a mix of research and philosophical deduction to argue the importance of developing a curriculum of giving and thus a new view of education. In this view, we realize that things like standardized testing, normative assessments and school league tables are only very limited, and often constricting, measures of quality education. If individual wellbeing and social cohesion are the horses that pull the cart of academic competencies, then it seems that governments around the world are trying to put the cart in front of the horses.

Schools that see the wellbeing of students to be the responsibility of all staff, not just the Chaplain or school counselor, have students with higher levels of wellbeing *and* academic achievement (Spratt et al. 2006; Lovat et al. 2009). Often I am puzzled about the fact that this doesn't seem to be rocket science, and yet testing and quantitative measuring seems more and more oppressive to teachers trying to focus on what really matters. What really matters is a whole person approach to education, in which academic success is seen as a by-product, however important. Learning how to wear leather sandals – how to give to self, others and life itself – seems to be an essential ingredient in becoming such a whole person.

Acknowledgements

Special thanks go to Steve McNab for insightful comments and editing, as well as to the teachers Sandra Griffin, Shannon Brown and Leonie Flynn who have contributed to the data in this chapter.

Sections of the research presented in this article have also received treatment in previous publications, most notably in Chapter 7 of Lovat, T., Toomey, R., Clement, N., Crotty, R. and Nielsen, T. (2009) *Values Education, Quality Teaching and Service Learning: A Troika for Effective Teaching and Teacher Training*, Australia: David Barlow Publishing

References

AGDEST (2008) *At the Heart of What We Do: Values Education at the Centre of Schooling – The Final Report of the Values Education Good Practice Schools Project – Stage 2.* Carlton South, Vic.: Curriculum Corporation.

Benninga, J. S., Berkowitz, M. W., Kuehn, P. and Smith, K. (2003) 'The relationship of character education implementation and academic achievement in elementary schools',

Journal of Research in Character Education, 1(1): 19–31.

Benninga, J. S., Berkowitz, M. W., Kuehn, P. and Smith, K. (2006) 'Character and academics: What good schools do', *Phi Delta Kappan*, 87(6): 448–52.

Berkowitz, M. W. and Bier, M. C. (2004) 'Research-based character education', *Annals of the American Academy of Political and Social Science*, 59(1): 72–85.

Berkowitz, M. W. and Bier, M. C. (2007) 'What works in character education', *Journal of Research in Character Education*, 5(1): 29.

Billig, S. H. (2000) 'Research on K-12 school-based service learning: The evidence builds', *Phi Delta Kappan*, 81(9): 658.

Billig, S. H. (2007) 'Unpacking what works in service-learning: Promising research-based practices to improve student outcomes', in J. Kielsmeier, M. Neal and N. Schultz (eds) *Growing Into Greatness 2007*, Saint Paul, MN: National Youth Leadership Council, 18–28.

Braun-Courville, D. K. and Rojas, M. (2009) 'Exposure to sexually explicit web sites and adolescent sexual attitudes and behaviors', *The Journal of Adolescent Health: Official Publication of the Society for Adolescent Medicine*, 45(2): 156–62, doi:10.1016/j.jadohealth.2008.12.004.

Brooks, K. (2008) *Consuming Innocence: Popular Culture and Our Children*, Brisbane: University of Queensland Press.

Caprara, G. V., Barbaranelli, C., Pastorelli, C., Bandura, A. and Zimbardo, P. G. (2000) 'Prosocial foundations of children's academic achievement', *Psychological Science*, 11(4): 302–6.

Childs, R., Thunderbox Production (Producers) and Stanley, F. (Director) (2008) *Our Kids at Risk*, Video/DVD, Future Makers.

Conrad, D. and Hedin, D. (1991) 'School-based community service: What we know from research and theory', *Phi Delta Kappan*, 72(10): 743–9.

Csikszentmihalyi, M. (2000) 'Happiness, flow, and economic equality' *The American Psychologist*, 55(10): 1163–4.

Csikszentmihalyi, M. (2002) 'The good work', *NAMTA Journal*, 27(3): 67–82, retrieved from ERIC database.

Deakin Crick, R., Coates, M., Taylor, M. J. and Ritchie, S. (2004) *A Systematic Review of the Impact of Citizenship Education on the Provision of Schooling*, London: EPPI-Centre, Social Science Research Unit, Institute of Education.

Deakin Crick, R., Taylor, M. J., Tew, M., Samuel, E., Durant, K. and Richie, S. (2005) 'A systematic review of the impact of citizenship education on student learning and achievement', *Research Evidence in Education Library*, London: EPPI-Centre, Social Science Research Unit, Institute of Education.

Dillon, M., Wink, P. and Fay, K. (2003) 'Is spirituality detrimental to generativity?', *Journal for the Scientific Study of Religion*, 42(3): 427–42, retrieved from CSA Sociological Abstracts database.

Dolan, P., Peasgood, T. and White, M. (2008) 'Do we really know what makes us happy? A review of the economic literature on the factors associated with subjective well-being', *Journal of Economic Psychology*, 29, 94–122.

Ewing, R.and Smith, D. (2003) 'Retaining quality beginning teachers in the profession', *English Teaching: Practice and Critique*, 2(1): 15–32.

Forgas, J. P. (2007) 'When sad is better than being happy: Negative affect can improve the quality and effectiveness of persuasive messages and social influence strategies', *Journal of Experimental Social Psychology*, 43, 513–28.

Grossman, D. and DeGaetano, G. (1999) *Stop Teaching Our Kids to Kill: A Call to Action*

Against TV, Movie & Video Game Violence, New York: Crown Publisher.

Hamzelou, J. (2010) 'Happiness ain't all it's cracked up to be', *NewScientist*, http://www.newscientist.com/article/dn18585-happiness-aint-all-its-cracked-up-to-be.html

Huston, A., Donnerstein, E., Fairchild, H., Feshbach, D., Katz, P., Murray, J., Rubinstein, E., Wilcox, B. and Zuckerman, D. (1992) *Big World, Small Screen; The Role of Television in American Society*, Lincoln, NE: University of Nebraska Press.

Isen, A. M., Rsonezweig, A. S. and Young, M. J. (1991) 'The influence of positive affect on clinical problem solving', *Medical Decision Making*, 11, 221–7.

Juvonen, J. and Wentzel, K. R. (1996) *Social Motivation: Understanding Children's School Adjustment*, New York: Cambridge University Press.

Lovat, T. J. and Toomey, R. (eds) (2007) *Values Education and Quality Teaching: The Double Helix Effect*, Terrigal, NSW: David Barlow Publishing.

Lovat, T., Toomey, R., Clement, N., Crotty, R. and Nielsen, T. (2009) *Values Education, Quality Teaching and Service Learning*, Terrigal, NSW: David Barlow Publishing.

Maslow, A. H. (1943) 'A theory of human motivation', *Psychological Review*, 50(4): 370–96.

Neff, K. D. and Vonk, R. (2009) 'Self-compassion versus global self-esteem: Two different ways of relating to oneself', *Journal of Personality*, 77(1): 23–50, doi:10.1111/j.1467-6494.2008.00537

Neff, K., Rude, S. and Kirkpatrick, K. (no date) 'Running head: Self-compassion and positive functioning', *Journal of Research in Personality.*

Nielsen, T. W. (2005) 'Values education through thinking, feeling and doing', *Social Educator*, 23 (2): 39–48.

Nielsen, T. W. (2010) 'Towards pedagogy of giving for wellbeing and social engagement', in Lovat, T. and Toomey, R. (eds) *International Research Handbook on Values Education and Student Well-being*, New York: Springer.

Oman, D. and Thoresen, C. E. (2000) 'Role of volunteering in health and happiness', *Career Planning and Adult Development Journal*, 15(4): 59–70, retrieved from ERIC database.

Post, S. and Neimark, J. (2007) *Why Good Things Happen to Good People*, New York: Broadway Books.

Scales, P. C., Roehlkepartain, E. C., Neal, M., Kielsmeier, J. C. and Benson, P. L. (2006) 'Reducing academic achievement gaps: The role of community service and service-learning', *The Journal of Experiential Education*, 29(1): 38.

Schnall, S., Jaswal, V. K. and Rowe, C. (2008) 'A hidden cost of happiness in children', *Developmental Science*, 11(5): 25–30.

Seligman, M. (2002a) *Authentic Happiness: Using the New Positive Psychology to Realize Your Potential for Lasting Fulfillment*, New York: Free Press.

Seligman, M. (2002b) 'Positive psychology, positive prevention, and positive therapy', in C. R. Snyder and S. J. Lopez (eds) *Handbook of Positive Psychology*, 3–9.

Seligman, M. (2008) 'Positive health', *Applied Psychology*, 57(1): 3–18.

Seligman, M. and Csikszentmihalyi, M. (2000) 'Positive psychology: An introduction', *American Psychologist*, 55(1): 5–14.

Spratt, J., Shucksmith, J., Philip, K. and Watson, C. (2006) '"Part of who we are as a school should include responsibility for well-being": Links between the school environment, mental health and behavior', *Pastoral Care in Education*, 24(3): 11–21.

Tolle, E. (2005) *The Power of Now*, London: Hodder and Stoughton.

Townsend, K. K. (1992) *Why Johnny Can't Tell Right From Wrong: The Most Important Lesson Our Schools Don't Teach – Values*, http://www.findarticles.com/p/articles/

mi_m1316/is_n12_v24/ai_13252028, accessed 8 February 2011.

Welsh, M., Parke, R. D., Widaman, K. and O'Neil, R. (2001) 'Linkages between children's social and academic competence: A longitudinal analysis', *Journal of School Psychology*, 39(6): 463–82.

Wentzel, K. R. (1991a) 'Relations between social competence and academic achievement in early adolescence', *Child Development*, 62, 1066–78.

Wentzel, K. R. (1991b) 'Social competence at school: Relation between social responsibility and academic achievement', *Review of Educational Research*, 61(1): 1–24.

Wentzel, K. R. (1993) 'Does being good make the grade? Social behavior and academic competence in middle school', *Journal of Educational Psychology*, 85, 357–64.

Wentzel, K. R. (1996) 'Motivation in context: Social relationships and achievement in middle school', in J. Juvonen and K. R. Wentzel (eds) *Social Motivation: Understanding Children's School Adjustment*, Cambridge and New York: Cambridge University Press, 226–47.

World Health Organization (2008) *Suicide Prevention*, http://www.who.int/mental_health/prevention/suicide/suicideprevent/en/, accessed 3 November 2008.

16

DEVELOPING WISDOM

The possibilities of a transformative education

Roslyn Arnold

What do I wish I had been told as I embarked on a career in education some decades ago? What might give hope, comfort and inspiration to the next generation of influential men and women? Does anything stay the same in the rapid whorl of ever-lasting change? What is emotionally, psychically and ecologically sustaining and re-vivifying in public and private life?

Five years ago at the end of my book, *Empathic Intelligence*, I prophesized that we were entering the dawn of a new age of enlightenment in which 'the discovered worlds will be closer to home than we have ever imagined: our own dynamic minds shaped by experience and engagement with others' (Arnold 2005: 224). Reading that now, my optimism is even greater that we could be entering a period of such compelling revelations into brain potential that possibilities for education of the mind could transform the field of education. Some dynamic leaders in faculties of education are recognizing these advances in studies of intellectual development. Supported by a solid body of scholarly research, they have developed cross-disciplinary courses that challenge old orthodoxies and offer students exciting, mind-provoking and transformative experiences. They trust that high quality engaging experiences with like-minded individuals will be powerfully formative in developing transformative practices to inform professional life.

No longer do we need to rely on assertion and faith to support claims about human potential. Evidence is now available to support what formerly might have been known only intuitively. One of the promising developments in science is the increasing interest in the nature and function of the emotions in humans. Under the influence of scientific research into human consciousness (Rose 1993, 1998; Damasio 1994, 2000, 2003; LeDoux 1996, 2003) attitudes to human thinking and the nature of intellectual maturity are changing. The concept of rational thought, associated particularly with the work of the seventeenth-century philosopher Rene Descartes, is being re-conceptualized since brain neural imaging and research on the development of consciousness illuminates the synergies between the cognitive and emotional parts of human brains (Sternberg 1990; Czikszentmihalyi 1990, 1998; LeDoux 1992, 1996, 2003; Damasio 1994, 2000, 2003; Eisenberg and Strayer 1996; Carpenter 2002; Davia 2002).

The work of Daniel Stern (1985) on the role of empathy in infancy illuminates the importance of empathic attunement in early learning, emotional development and socialization. Stern writes about the development of 'intersubjective relatedness', that ability to experience one's self as a separate being from others, but as a dependent being too. The process by which the mother's empathic responsiveness evokes, stimulates, validates and maybe names the infant's emotional and physical state, ensures that her underlying affective response is encoded in the baby's brain.

According to Stern, the degree to which the major affect states (interest, joy, surprise, anger, distress, fear, contempt, disgust and shame) are encoded in the baby's brain influences the development of their core relatedness. That sense of core relatedness is the basis for the development of inter-subjective relatedness. Stern argues that in the pre-verbal stage infants seek to share joint attention, intentions and affect states with significant others. The process of organizing the affect responses of the mother to the infant, and indeed, possibly, the infant's own physiological responses also, Stern refers to as the laying down of templates into 'Representations of Interactions that have been Generalised' (RIGS) (Stern 1985: 97). It is the mother's (and others') empathic responsiveness to the infant that influences the integrations of 'agency, coherence, and affectivity' to provide the infant with a unified sense of a core self and a core other: 'the existential bedrock of interpersonal relations' (Stern 1985: 125).

This substantial body of scientific research on brain and mind development along with the work of Noddings (1992, 1993, 1998) and Nussbaum (1995, 1997) on the function of care and ethics in human and moral development should encourage educators to aspire to transformative pedagogy. In that aspiration, educators are encouraged to pay more attention to the nature and quality of their relationships with students, seeking to understand deeply how inter-subjective and intra-subjective processes enhance transformative teaching and learning experiences. Part of the search for understanding can be informed by identifying the characteristics of the enabling contexts and relationships within which both teachers and students become open to possibilities beyond their expectations and reflective about them.

It is inherent in life that things change constantly. Reflective people monitor closely their own internal dynamics. For them, it is second nature to observe, reflect and analyze their own thoughts and feelings, and their interactions with others and the environment. Such people tend to develop a rich inner life – the product of that recursive, dynamic process of inter-subjective and intra-subjective engagements. Dynamism as it is theorized in empathic intelligence (Arnold 1995) refers to the sense of energy, tension or movement present when we relate with deep thought and feeling to a situation. Empathic attunement to self, others and the world of experience underpin that process. It is a psychic and recursive energy that moves outwards, and inwards propelled by the human capacity to learn from experience. Interpersonal relationships, work and play can involve dynamic experiences, alongside energy generated by our own mental and physical states.

Dynamism functions to fuel deep learning, insight and psychic development. Even within the concept of dynamism as outlined here, there is a role for stillness, and for attuned listening to self and to others. Dynamism in the case of poise or stillness might be thought of metaphorically as potential, rather than actual, energy.

Teachers skilled in attuning to the needs of others can recognize this internal, suspended energy and use it to fuel the psychic activity of reflecting upon both their thoughts and the feelings prior to responding to the other. In responding, they might choose to amplify the other's feelings or suggest a strategy. A common statement made by students' in classrooms is 'I hate maths'. A teacher can reply, 'You will need maths when you grow up so you must persevere' or 'It's frustrating when you can't quite remember the formula, isn't it? Let's go through the steps again.' It is easy to see which response is likely to mobilize the student. In their interactions with students, teachers have multiple opportunities for judicious choices in empathic responses.

Educators aspiring to be transformative need the professional expertise to mirror others, to attune to them and to be self-reflective. While empathic intelligence attempts to capture something of the resonance of human engagements, it acknowledges the necessity for participants themselves to observe, feel, intuit, think, introspect, imagine and test their own data gathering of phenomenological moments, mindful that in adopting a stance of engaged and subjective objectivity, that attitude itself will influence the phenomena of engagement. That is, the nature of an engagement is influenced by all the dynamics in play. Phenomenological moments arise without pre-warning but often as the result of careful attunement, preparation and sensitivity to dynamics. They are moments of high emotional impact that result in a significant shift in thinking and awareness. They can be pleasant (surprising, exciting, exhilarating) or unpleasant (shocking, distressing), but their lasting impact upon thinking and feeling depends very much on how well they are understood and integrated by individuals.

Of particular importance in mobilizing positive effects in students is the function of hope. It is an emotion that exerts a powerful effect on students' capacity to mobilize coping and learning strategies. Hope has a natural affinity with empathic intelligence because it functions best as both a quality of feeling and of thinking. In its feeling capacity it acknowledges the depth of another's aspirations, disappointments and efforts; in its thinking capacity it recognizes the function of denial, the value of introspection and insight, and the limits of rationality in mobilizing change. In its integrated capacity synthesizing both feeling and thought, it conveys a belief in the endurability of human growth, the persistence of psychic evolution and belief in the tacit abilities of humans (Polanyi 1983) to mobilize internal resources in the service of survival and the search for meaning.

Transformative pedagogy lightens up the brain and fires the mind to reach for the ineffable. Erudite scientific papers, popular books and lifestyle magazines carry messages about the possibilities of the human brain to work more productively and efficiently throughout a lifetime. The scholarly community in education and its related disciplines is hearing the message but are policy makers keeping up?

With the promise of a National Curriculum in Australia on the near horizon, those of an optimistic disposition might have hoped that at last here was an irresistible opportunity to free school syllabuses from the tyranny of overcrowded content-loaded curricula. Now is the time to favour curricula rich in choices for in-depth study of content through enriching experiences. It is doubtful whether that optimism is justified. It is a sad irony that at a time when the world is replete with masses of information on every subject known to humanity, readily accessible to even very young children, we are over-relying on teachers in schools to roll out information as if knowing facts was the measure of education. The real challenge of a good education is to guide students to make sound, informed judgments about the worth of the information they can source from multiple repositories. Are institutions of learning across all spectra providing that curricula? Are students graduating fired with enthusiasm to continue their education throughout life and work? Adaptation, learning and the 'getting of wisdom' can be decades-long pursuits. In formal education settings, how can educators create enabling conditions for transformative pedagogy? What is it?

I postulate that transformative pedagogy is so powerfully enlightening and affectively charged that the mind experiencing it reaches an insight from which there is no lasting regression. This is not to say there may not be a reduction in the affective strength of the experience, or even its emotional memory when the experience is recalled, or when something like it is repeated, rather to say that having known such moments, one is confident that there may be more such. In transformative experiences, pathways of possibility have been opened in the mind and with care and confidence those pathways remain accessible to further possibilities. In an iterative process whereby the remembrance of past transformative experiences and hope co-operate to generate confidence and expectation, new learning, that might otherwise be unremarkable, becomes significant. An example might be the performance of a music piece in which the player discovers a nuance of interpretation that is exciting and revelatory and realizes that the performance of other pieces offers similar interpretative possibilities. High-order abilities of thinking, feeling, experimenting and creating, combined with talent, underpin the transformative experience in this case.

When I write poetry and discover a metaphor or rhythm that symbolizes and communicates to me the cohesion of thought and feeling particular to the experience I am formulating, I experience transformative moments. What might have been observed, felt and thought in fragmentary and elusive ways is now coherent, patterned and energized. I have created a personal, psychic pathway in my mind. When I re-read such poetry, the original experience is reactivated in a nuanced and energizing way. In a complex mood of wonder and despair, the English poet Gerard Manley Hopkins (1918) captured something of the challenges of creative explorations into the human condition:

> O the mind, mind has mountains: cliffs of fall
> Frightful, sheer, no-man-fathomed

Transformative experiences are the outcome of courage, confrontations with the ineffable and confidence in the worth of the quest. Teachers aware of the risks, the rewards and the necessary preparations can make ideal mentors.

Transformative pedagogy seeks more complex outcomes than a search for knowledge of the bounded, quantifiable kind. It refers to pedagogy that engages thinking, questing, speculating, puzzling and feeling across a range of emotions in a search of understanding, insight, even wisdom. It draws on tacit and deeply internalized memories of often lifelong learning experiences, building on those that are positive and generative and bringing to the fore those that are inhibiting. Because this approach to pedagogy is complex and sometimes opaque in nature, often it requires empathy, dialogue, introspection, reflection, research, modeling, testing, hypothesizing, reformulating, practice and patience before the mind structures the dynamics of thought and feeling into the transformative power of revelatory insights. In preparation, teachers need to help students to access information and evaluate it, but most of all they need to be able to identify how such information serves students' needs and enhances their formative experiences of maturing to adult life.

While the case is yet to be made for a relationship between transformative pedagogy and the development of wisdom, many of the characteristics of wisdom identified by a number of scholars in the field (see Sternberg 1990 for a comprehensive introduction to the field) are associated with teachers who enact and model the qualities of empathic intelligence (Arnold 2005), especially in demonstrating ability to work with expertise, engagement, enthusiasm and intelligent care (Arnold 2005) in the service of others.

As with any complex concept, definitions of wisdom differ in emphasis but include characteristics such as:

- A recognition of the presence of experiences in life that are emotionally, mentally and ethically challenging.
- Breadth and depth of understanding.
- Recognition that knowledge is uncertain or provisional and truth is not absolutely knowable.
- Willingness and exceptional ability to formulate sound, practical judgments in the face of life's uncertainties.
- Ability to transcend personal needs, thoughts and feelings.

(Selected and adapted from Sternberg, 1990)

I would add:

- Ability to recognize and tolerate ambiguity.
- Insatiable curiosity.
- Ability to dwell in imponderables.
- A predisposition to introspect spiritually.
- Ability to access and integrate affect and cognition.
- Aspiration to achieve insight.

These characteristics allude to the nature of wisdom. It is highly likely that when talking about wisdom or contemplating its essential qualities, one looks up and out rather than down. It has that kind of effect because it entices one to look beyond or deeply inward.

The development of wisdom with its high-level characteristics of intellect, personal morality and professional efficacy is supported by qualities such as resilience, risk taking, problem solving, pattern recognition, the application of multiple intelligences, habits of deep reflection, theory building, theory testing and speculative thinking. A mind working in that kind of mental playground is repeatedly observing, rehearsing, experimenting, creating, imagining and growing. Seemingly, the brain thrives on challenge and can grow new pathways in response to stimulation. It can grow till the last heartbeat and thrives in socially, emotionally engaging environments.

While it is a very challenging aspiration to educate for wisdom, clearly certain life experiences provide pathways to such education. Possibly, given the often uninspiring objectives of much school and tertiary education, it happens nonetheless under the guidance of mentors and teachers with vision. Probably such teachers have to be somewhat subversive and courageous in looking beyond mandated curricula or standards to challenge convention and to model creative ways of knowing and being. Recently I observed a class of Year 12 girls responding with engagement and insight to a teacher-led discussion of Shakespeare's *Richard III*. The line of questioning was around the relevance of the play. One student objected to that. With passion and conviction she asked rhetorically why that question was even considered since the experience of reading Shakespeare was so personally and aesthetically rewarding. The experience, she asserted, went far beyond anything one could put into words. From the radiant look on her face and the tenor of her voice it was clear that Shakespeare had worked his magic on her. The rest of us just watched and listened with awe. Transformation moments are powerfully affecting to witness. What life experiences had set her on the path to wisdom? Will formal education continue to enhance her or kill her spirit?

In reflecting on this experience, I realized that as a teacher the moments that transformed my pedagogy in my early career were those where I acted somewhat subversively. These were moments of sheer inspiration when in response to my frustration at the meaninglessness of tasks that my students or I were supposed to do and with a subliminal awareness that something better was possible, I took some subversive action. That action actually liberated me and opened up my students to possibilities. For example, when instructed to fill in the report cards for a class whom I had only just met, I asked each student to tell me what he or she would like to have written against English in the report card. As the previous teacher had left without completing the duty, I had to do so. I had been warned that the class was very difficult and the reason for the previous teacher's sudden departure. Since I had not yet experienced the students as difficult I was not prepared to label them as such. Needless to say, each student had an idea of the desirable qualities of a good English student. With rather more enthusiasm than expertise, they dictated the comments they wanted sent home. With a bit of editing I wrote their instructions directly into the spaces. The reports were sent home.

I didn't find that class difficult at all. One young lad did tell me later that his father was giving him a bike for Christmas, as he had never had a good comment before. Decades later I still count that as a transformative moment in my development as a teacher. I am happy to counter any objections to the strategy. Oddly enough, the action was not astonishingly brilliant, nor especially subversive, just a shift in focus to the needs of the students rather than the needs of someone or something well outside the classroom. It was an irresistible opportunity. It gave me the confidence to search for more such opportunities. They can and should be mandated as pedagogy but they can carry personal risk for teachers unless they work in an affirming environment where aspiration goals are promulgated with integrity.

Another transformative moment was when I was engaged in a longitudinal study of school children's writing abilities for my PhD research. I was teaching writing to a group of students in two different schools, using letter writing between peers. Each student had an unknown pen friend in the partner school. Occasionally, when students had run out of ideas to write about, I would suggest they comment to their pen friends on the writing program. In a letter to his writing partner, one boy wrote, 'We have a writing teacher comes to this school. She doesn't seem to teach us much but (that doesn't matter) … most of the magic is in the brain.' Inspired by the aptness of his metaphor, I wanted to use the phrase 'magic in the brain' as the title of a book I wrote on writing development. I was advised, possibly astutely, that such a book risked being catalogued under 'The Occult'. While these days a book on the brain might have its own section in a bookshop, magic might struggle for acceptance.

It matters little whether I am right or wrong in prophesying a new age of enlightenment in understanding the possibilities of human development. What matters most is the hope and aspiration generated by that prophesy. Powerful feelings and penetrating thought generated and tuned through observation and introspection then shared with a community of like-minded others contribute something to the ever-sustainable world of imagination and ideas.

References

Arnold, R. (2005) *Empathic Intelligence: Teaching, Learning, Relating*, Sydney: University of New South Wales Press.

Carpenter, P. (2002) 'Evidence of perception, cognition and individual differences', unpublished paper presented at the *Learning and the Brain Conference*, Boston, MA.

Csikszentmihalyi, M. (1990) *Flow: The Psychology of Optimal Experience*, New York: Harper and Row.

Csikszentmihalyi, M. (1998) *Finding Flow: The Psychology of Engagement with Everyday Life*, New York: Basic Books.

Damasio, A. (1994) *Descartes' Error: Emotion, Reason and the Human Brain*, New York: Grosset/Putnam.

Damasio, A. (2000) *The Feeling of What Happens: Body, Emotion and the Making of Consciousness*, London: Vintage.

Damasio, A. (2003) *Looking for Spinoza: Joy, Sorrow and the Feeling Brain*, Orlando, FL: Harcourt Books.

Davia, C. (2002) 'Minds, brains, chaos and catalysis: An ontological approach to the mind/brain problem', unpublished paper presented at the *Learning and the Brain Conference*, Boston, MA.

Eisenberg, N. and Strayer, J. (eds) (1996) *Empathy and its Development*, Cambridge: Cambridge University Press.

Hopkins, G. (1918) 'No worst, there is none. Pitched past pitch of grief', http://www.poemhunter.com/Gerard Manley Hopkins, accessed 22 November 2010.

LeDoux, J. (1992) 'Emotion and the amygdala', in Agglington, A.P. (ed.) *The Amygdala: Neurobiological Aspects of Emotion, Memory and Emotional Dysfunction*, New York: Wiley-Liss.

LeDoux, J. (1996) *The Emotional Brain: The Mysterious Underpinnings of Emotional Life*, New York: Simon & Schuster.

LeDoux, J. (2003) *Synaptic Self: How Our Brains Become Who We Are*, New York: Penguin Books.

Noddings, N. (1992) *The Challenge to Care in Schools: An Alternative Approach to Education*, Columbia University, NY: Teachers College Press.

Noddings, N. (1993) *Educating for Intelligent Belief or Unbelief*, Columbia University, NY: Teachers College Press.

Noddings, N. (1998) 'Ethics and the imagination', in Ayers, W. and Miller, J.L. (eds) *A Light in Dark Times and the Unfinished Conversation: Maxine Greene*, Columbia University, NY: Teachers College Press.

Nussbaum, M. (1995) *Poetic Justice: The Literary Imagination and Public Life*, Boston, MA: Beacon Press.

Nussbaum, M. (1997) *Cultivating Humanity: A Classical Defense of Reform in Liberal Education*, Cambridge, MA: Harvard University Press.

Polanyi, M. (1983) *The Tacit Dimension*, Gloucester, MA: Peter Smith.

Rose, S. (1993) *The Making of Memory*, Toronto: Bantam Books.

Rose, S. (ed.) (1998) *From Brains to Consciousness: Essays on the New Science of the Mind*, Princeton, NJ: Princeton University Press.

Stern, D. (1985) *The Interpersonal World of the Infant*, New York: Basic Books.

Sternberg, R. (ed.) (1990) *Wisdom: Its Nature, Origins and Development*, Cambridge: Cambridge University Press.

17

THE SCHOOL OF WORLD PEACE

Robin Grille

My early twenties were troubled times for me. In school I had learned much about the world – but diddlysquat about myself. I was rudderless, hopping randomly from one job to another in a haphazard search for purpose.

I was not alone in this anguish, as I was to discover later, when I eventually found my path in the field of psychology (with the guidance of non-academic mentors). In private practice I saw that so much of what passes for success in our culture masks a bottomless well of disillusionment. What happens to the legions of the bored, the trapped and the unfulfilled? Job discontent incubates all manner of social ills: depression, anger, addiction and discord – entire families feel the ripple effects. So many people detest their work, living for scraps of pleasure on weekends, accumulating goods and trinkets to compensate for inner emptiness. Today, this desperate overconsumption is what turns the wheels of our economy – while despoiling our earth.

When I ask my clients if their schoolteachers ever asked them, as children, what they really loved, almost all say no. Certainly, we were all offered elective subjects, though that is not until quite late, usually by the middle of high school. But precious few of us had teachers who showed an interest in and commitment to our personal passions. I find this shattering. How many millions of us spend our lives corralled into a career that others chose for us?

Besides the wretchedness of performing a loveless task day in, day out, there is an even bigger price we all pay for an education system that neglects children's unique and diverse passions. Countless individuals feel devalued in their school system; they are herded into activities for which they have no love, then subjected to the public humiliation of grading and ranking. As we will see later, this marginalisation can lead to reactive violence – at huge social cost. Education drives social evolution, and must surely carry some of the burden of responsibility for producing a society that is ruinously unsustainable, with epidemic levels of depression and growing violence. Here I wish to discuss innovative styles of education that have yielded remarkable reductions in social violence while elevating levels of student well-being and fulfilment beyond the familiar.

Even as a primary school boy, it struck me as ironic that, in a supposedly democratic society where self-motivation is prized, every classroom bore the hallmarks of a military academy where learning was predicated on obedience. How does anyone learn to voice their feelings in a do-as-you're-told environment? This flagrant contradiction remained with me viscerally and compelled me, decades hence, to send my daughter to a progressive primary school in Sydney. The search for this rare gem: a school that would trust children to follow their passions led me on a fascinating journey of discovery. The unexpected surprise was to find that child-centred education not only enhances learning and makes children happier, it also sharply reduces school violence. This finding is so consistent internationally, that I now believe child-led education to be a blueprint for reducing violence around the world.

The shape of future civilisations is determined by the way our children's socio-emotional needs are met, from early childhood through adolescence. In other words, we can educate for the status quo, or we can educate for an evolution towards a more loving, peaceful and sustainable society. A grandiose and sweeping statement, perhaps – were it not justifiable in light of new and revolutionary understandings about the human brain.

Recent advancements in neuropsychology have assigned unprecedented credibility to the notion that childhood drives world affairs. Our dozen or so years in school do far more than influence our thinking and fill our heads with data. For better or for worse, schooling leaves a deep imprint in the emotional and relational centres of each student's brain. In childhood and adolescence, the human brain is subjected to profound chemical and synaptic changes wrought through the impact of human relationships. These changes underpin the formation of individual personality and relating styles: the building blocks of any society.

Whether in the classroom, the playground or at home, an unrelentingly stressful environment damages the developing brain. Daniel Siegel, psychiatrist and leading world authority in the new field of interpersonal neurobiology explains it this way: 'The release of stress hormones leads to the excessive death of neurons in the crucial pathways involving the neo-cortex and limbic system – the areas responsible for emotional regulation' (Siegel 1996: 106). A child in fear who is not soon enough comforted loses brain cells (millions of brain cells if the fear is severe enough) in the hippocampus and orbitofrontal cortex: brain areas responsible for human empathy and impulse control.

The human propensity for unrestrained self-interest is not a given, as Milton Friedman would have us believe. The unbridled narcissism driving modern economic orthodoxy is a *symptom*, not a genetic imperative for our species. Friedman's assumption about the primacy of self-interest is now found to be false. Greed, like violence, has been exposed as the result of brain damage – an imbalance that is categorically avoidable and remediable (Perry 1997; Lewis et al. 2000; Gerhardt 2004).

Children who are made to feel ashamed, or intimidated by their teachers, are in fact hemorrhaging their potential. And yet, shame and fear are commonplace in schools where children are regularly set up to compete against each other and be subjected to the indignity of grading and ranking. The mainstream and most familiar model of schooling around the world is inherently authoritarian, disconnected from the child's personal passions, and remorselessly competitive. Rarely are children asked about their unique loves; instead they are pushed to perform under agenda imposed by adults who barely know them. This kind of schooling is a high-cortisol (stress hormone) environment, hardly the place where children will learn best about contributing to the world creatively, joyfully, with love and with compassion.

When a child is shamed, either by a teacher or by the disgrace of a poor exam result, the resulting biochemistry of stress dowses self-confidence and potentially triggers a simmering, reactive hostility (Schore 1994; Gerhardt 2004). Shame or humiliation spark off a chain reaction in the child's brain that suppresses the ability for empathy or attention to others. The emotionally charged amygdala (centre of emotional processing) overwhelms the prefrontal cortex, so the child's ability to attend to others is drowned by self-focused anxiety. Shame erodes the human capacity for empathy (Goleman 2006). A repeatedly intimidated or shamed child is at risk of becoming a detached, aggressive or selfish adult.

Do fear and shame offer any educational advantage in return for what they rob from a child? Hardly – fear and shame can only harm academic potential. Elevated cortisol interferes with learning and memory retention.

To latter-day neurobiologists, a new and more harmonious society begins with the brain chemistry of amity and collaboration. Human affection produces a biochemical cocktail that generates growth in the brain's empathy centres (Gerhardt 2004). Repeated exposure to oxytocin, the potion dubbed as the love hormone, quite literally builds a more empathic brain. Oxytocin doubles up as a nutrient that seeds synaptic growth. Warm relationships are the key generators of oxytocin. Affection and compassion in the classroom are no longer the stuff of sentimentality, they are not a warm-fuzzy luxury – these are critical biological requirements if our children are to mature into the fullness of their true potential and if societies are to evolve towards peace and sustainability.

The age of Cartesian rationalism is over: it is confirmed that we are creatures of emotion and relationship. The dominance of emotional regions of the brain is a verifiable property of the human neural circuitry (Lewis et al. 2000). The discovery that the limbic (emotional) brain overrules the coolly analytical forebrain adds urgency to the growing realisation that emotional intelligence should be uppermost in the school agenda. Education needs a greater focus on love and on human relations, and an end to authoritarian, top-down styles of delivery.

Authoritarian pedagogy (the unilateral, 'because I said so' approach) comes at a great cost to societies – it hardens attitudes, poisons democracy and increases violence. This causal link is as empirical as it is intuitive. Social researchers have confirmed what brain scientists would predict: that individuals who were brought up under punitive parenting styles tend to favour harsher law-and-order policies domestically, they are more likely to support war as a foreign policy, and they are more likely to oppose pro-environment policies (Milburn and Conrad 1996). Certainly, there are always individual variances and exceptions – but en masse, conservative governments with an authoritarian penchant follow from authoritarian pedagogy. Coercive education conspires with punitive parenting to guarantee the most conservative politicians their next generation of voters.

Psychohistorians have long understood that the way children are treated will write the destiny of any society. Let's examine two modern nations, their prevailing child-rearing and educational trends and their socio-political sequelae – this will highlight why childhood and education deserve to be the greatest priority for anyone concerned with social evolution.

The Pew Research Center, a pollster of global attitudes and trends, found recently that opinions about the USA had soured widely: 'In the view of much of the world, the United States has played the role of bully in the school yard, throwing its weight around with little regard for others' interests'.[1]

What kinds of acts is this 'bully' charged with? As the nerve centre of free-market fundamentalism, the American deregulatory free-for-all is widely understood to be a major factor behind the global financial crisis of 2009. In 2000, the average American CEO remuneration topped 500 times the *average* wage, and by 2005 it had rocketed to 800 times the *minimum* wage (Mishel 2006). Contrast this with the British CEO-to-average-wage ratio of 22:1 in the year 2005, Canada's ratio of 20:1, France's 15:1 and Japan's 11:1 (Prosser 2009).

On the environmental front, the USA has shown contempt for the international consensus of climate scientists who warn us of imminent ecological calamity – it has stubbornly refused to ratify the Kyoto Protocol and failed to show leadership at the 2009 Copenhagen Summit.

The USA has the highest per capita prison population rate in the world, has scandalised the world with its use of torture as a military strategy and has dismantled habeas corpus and other civil liberties. The Supreme Court in January 2010 removed caps on political 'donations' from big business: another nail in the coffin of democracy (Liptak 2010).

Ironically, the Pew report's 'school-bully' metaphor hints strongly at the root cause of America's fall from grace. The main clues to the hawkish and increasingly undemocratic character of American foreign and domestic policy lie in American family and education policies. Corporal punishment in schools is still legal in 20 states, where teachers strike children on the buttocks with a wooden paddle.[2] The USA is the only developed nation to cling to this anachronistic practice. In public schools, a national 'zero tolerance' policy is increasingly morphing schools into prison-like institutions, patrolled by armed guards, where discipline is meted out by police, where students are surveyed by security cameras, handcuffed, searched, arrested and suspended for the most trivial infractions. A staggering 97 percent of school suspensions involve infractions not related to the weapons, drugs or alcohol that this policy was purported to control. No empirical data indicates that the police-state approach to 'discipline' has reduced violence in schools. Meanwhile, a pharmaceutical approach to discipline means that Americans gobble 90 percent of the world's Ritalin, and four million American children are medicated for their behaviour (Soling 2009).

Troubled American teenagers are sent to unregulated, privately owned juvenile boot camps, where they face physical beatings and solitary confinement. Several children have lost their lives in these camps.[3]

The USA is one of two remaining nations (along with Somalia) not to have ratified the UN Convention on the Rights of the Child.[4] Little wonder that the USA earned second last place in a 2007 United Nations Children's Fund (UNICEF) report on childhood well-being in affluent nations.[5]

Second from the top of the same UNICEF list is Sweden,[6] a nation that boasts one of the lowest homicide rates in the world and a commitment to 100 percent independence from fossil fuels by 2020.[7] Sweden was the first country to ban corporal punishment in schools *and* homes as of 1979. Its generous paid parental leave provisions, high breastfeeding rates (supported by a raft of government and World Health Organization (WHO) initiatives) and open preschools have combined to produce one of the world's fairest, most equitable and sustainable societies (Grille 2005: 162–3).

Today we can be more certain than ever that child-rearing reform and educational reform are the most powerful strategies for achieving more harmonious societies. Armed with this know-how there is nothing to prevent us from transforming societies all around the globe, evolving well beyond what the Swedes have achieved at millennium's dawn. If you think that this knowledge should provoke a profound overhaul of educational approaches, there is good news in abundance. Let's take a glimpse at some of the most exciting new initiatives that are transforming classrooms, and children's behaviour, around the world.

At this point in the history of progressive education there are still few studies evaluating the link between child-centred educational methods and improved social well-being, reduced violence and better relationships. Much of the evidence is anecdotal, based on educators' self-reports, and more research is warranted. Nevertheless, a fascinating pattern is emerging around the world, showing a strong trend for non-coercive education to produce more harmonious school communities. Let's take a look at what can happen when children's emotional needs are addressed in the classroom.

When Australian psychologists, Robyn Dolby, Belinda Swan and Judy Croll (2004) trialled their attachment-based program in one of Sydney's most trouble-prone preschools, they were not prepared for the results. Swan and her team trained the staff to be highly tolerant of the full range of children's emotions, to be comforting, empathic and dependable listeners (the framework used is based on the Circle of Security model (Cooper et al. 2005). Pitting the power of love against the most stringent of tests, the psychologists chose a preschool in the heartland of Sydney's disadvantaged: plagued by domestic violence, family fragmentation and substance abuse. The behavioural incident logbook at this school was off the scale, driving dispirited staff towards fortnightly turnover rates. By prioritising the children's emotionally charged communication, the school's culture was overhauled – children's challenging behaviours and staff turnover rates both plummeted (Dolby et al. 2004). Moved by these unexpectedly impressive results, the Benevolent Society has replicated and delivered this program, based on the Attachment Matters model, to fifteen childcare centres and preschools. In every case there have been significant reductions in conduct and peer problems and a discernible increase in pro-social behavior. Staff morale, of course, is also blossoming at these centres. Is this not a promising blueprint for societal change? If profound changes were produced among such troubled children in the classroom, then what would be the potential for society at large if most children were educated in this way?

A long-time teacher in progressive education settings, Juli Gassner from Kinma primary school,[8] Sydney (until recently, my daughter's school) will often allow her students to 'be the teacher', and to teach the rest of the class something that is their passion or area of expertise (examples have included: how to ride a skateboard, all about dinosaurs, etc.). Juli's students soon discover that any attempts to be controlling or pushy will invite dissent, alienate others and spoil co-operation. Authoritarianism quickly loses any appeal when students don the teacher's shoes for a while. As Juli explains: 'it is only by having more responsibility that children gain a sense that they are responsible for their consequences'. In Kinma classrooms significant time is devoted to supporting children to resolve their conflicts together, as a group. Every child is invited to voice his or her feelings. They do not teach *about* democracy here – the children directly and intimately partake of the 'creative chaos' of democracy. Juli maintains that the more children learn a language for expressing emotion freely, responsibly and safely, the less they act out their frustrations destructively against each other, the teachers or school property. It does not suffice to *teach* this vital social skill – it needs to be practiced, as we do with multiplication tables, over and over in the classroom. What kind of world would we have if this were an educational priority rather than a luxury?

Around Australia, a growing number of schools are adopting 'emergent' curricula, democratic practices and restorative justice (the latter is a dialogical method for enhancing understanding and making peace between bullies and victims) – all measures that improve emotional well-being and enhance relationships greatly.

The restorative justice practice involves rejecting the old paradigm of blame and punishment in favour of dialogue: inviting children – both 'bully' and 'victim' – to express their needs and feelings. The success of this system as a genuine peacemaker is evidenced in its rapid proliferation around the world, in nations including Brazil, UK, USA and Australia.[9]

The new-wave 'emergent curriculum' explicitly builds on interests expressed by students, rather than content goals imposed by teachers, industrialists or politicians. The children's ideas constitute one of the most important (though certainly not exclusive) sources of curriculum (Jones and Nimmo 1994). In other words, the curriculum serves the child, rather than the other way around.

Sydney's Catholic pre- and primary schools are revolutionising their classrooms with the collaboration of the Institute for Early Childhood (IEC),

Macquarie University (O'Connor 2010). The project involves a transition to enquiry-based, collaborative learning. It is a shift in pedagogical practice that gives children time to explore their own interests. Kate O'Brien, Head of Primary Curriculum, says: 'research shows that children learn best from each other'. According to this evidence base they have built a collaborative culture – with marked improvements in social skills. Say, for example, some boys show a persistent interest in dragons. In a traditional setting they would have been asked to leave their dragon-lore outside – here their passion is embraced and in fact becomes the forum for engaging the required syllabus (reading about and researching dragons together, writing about dragons, measuring dragons, etc.). The teachers encourage the children's passion, rather than employing coercion to drive the learning.

To date, thirty-nine primary schools in Sydney have joined this project. A longitudinal evaluation (with the Institute of Early Childhood at Macquarie University) has yielded impressive data. Children's academic results had improved, particularly in relation to literacy and reading. Additionally, the children were more engaged in learning, worked more independently and stayed on task for longer periods. What is even more remarkable is that the children's *social* skills had improved, with significant decreases in negative behaviour and more evidence of pro-social behavior in both classroom and playground.[10] Why? Catholic Education Officer Franceyn O'Connor explains it this way: 'the children are interested in what they are learning and they feel valued', therefore: 'if you have engagement, you don't have behaviour problems'. None of this would surprise a neuropsychologist, who would have anticipated the results of flooding a classroom with oxytocin. What's more, this method is bearing academic fruit. Because the children are supported to pursue their passions, says Franceyn: 'they go *way* beyond the expected syllabus outcomes'.

A comparative study was set up to evaluate the performance of progressive British schools vis-à-vis the traditional and authoritarian school systems. The study looked at schools that embraced practices such as: student participation in school administrative decision making processes, creative and negotiative pedagogy and extensive extra-curricular activities over which the students had a significant degree of control (Hannam 2001).

The overwhelming self-report of head teachers and other senior managers was that 'student participation' impacts beneficially on self-esteem, motivation, sense of ownership and empowerment, and that this in turn enhances attainment. What's more, the study also identified academic advantages. In the more democratic schools, attendance rates were slightly higher, as were the levels of attainment at GCSE (Hannam 2001).

Beyond the academic advantages, the realisation that giving children a say in their own education yields a peace dividend is mushrooming around the world. Democratic educational trends are sprouting independently – as if humanity's collective unconscious is awakening to this idea all around.

One of the better-known examples of emergent or negotiated curricula is that of the Montessori system, of which there are over 5,000 schools in USA alone. The Montessori system has many unique attributes but of special interest here is its emphasis on student-chosen work, on collaboration, its refutation of grading and testing, and its instruction in both academic and social skills.

In a comparative study that tested children in hypothetical situations, Montessori children were significantly more likely (43 percent versus 18 percent of the control group from mainstream education systems) to use higher level reasoning rather than aggression to resolve conflicts. This trend was replicated in the field: 'Observations at the play-ground during recess indicated Montessori children were significantly more likely to be involved in positive shared peer play and significantly less likely to be involved in rough play that was ambiguous in intent ...' (Lillard and Else-Quest 2006). Montessori twelve year olds were also found to be 'significantly more likely to choose the positive assertive response'.

What has child-centred education achieved elsewhere? Japan is in the grip of an epidemic of youth violence. *Ijime*, school bullying, is out of control. One study by a children's research institute found that as many as 30 percent of high school and middle school students had experienced sudden acts of rage – *kireru* – at least once a month. Japan has also experienced a rising tide of serious youth crime, including arson, assault, rape, manslaughter and premeditated murder (Faiola 2004). This surge of youth violence may have many causes, but a major culprit is the strict, pressure-cooker approach that prevails in mainstream Japanese schools (Fredman 1995, Dogakinai 2010). Tens of thousands of Japanese children, collapsing under the pressure, break down and refuse to go to school at all. School refusal – *toko kyohi* – has multiplied tenfold in thirty years (Gordenker 2002).

The 'Free School' system in Japan was created as a response to the heavy-handedness of the mainstream and the widespread violence and dysfunction it has bred. These 'Free Schools' proved to be far more than a safe-haven alternative for the hordes of 'school refusers'. The students are free to choose their learning pathways according to their interests, and they can influence when, where and how they learn. This system has been so successful that the Japan Free School Association now boasts 240 member schools. Of particular interest to the social ecologist is the fact that this more gentle and liberal system is claimed to have reduced bullying among students (Gordenker 2002).

From the deeply troubled Middle East hails the voice of Yaacov Hecht – director of the Institute for Democratic Education in Tel Aviv – whose words bear an uncanny resemblance to Franceyn O'Connor's though the two have never met. Hecht elucidates the link between democratic education and peace thus: 'when kids are helped to do what they are passionate about, they are simply not interested in bullying anyone. They are too busy following their hearts. In such an environment they feel valued and special, and so the urge to violence disappears'. Israel is a veritable laboratory for democratic education; it now has twenty-five democratic schools, a democratic teachers' academy in Tel Aviv, and hundreds more schools in the process of conversion.[11] The prototype democratic school in Hadera received the National Education Prize in 1994.

Democratic education in Israel mirrors the nouveau 'emergent curriculum' that is proliferating in the West, and it is enjoying a rapid expansion worldwide. Each year the International Democratic Education Conference (IDEC) is attended by hundreds of delegates from many diverse nations.[12] This trend may have arrived not a moment too soon – Israeli democratic schools are particularly respected for their low incidence of violence. In 1998, Hecht and his team were asked to democratise the troubled Rogozin High School in a disadvantaged sector of Tel Aviv. By 2004, the school had become so attractive its population trebled. Previously a grim, prison-like campus, the school was now covered in murals, and it housed a local newspaper, a TV production studio and a clowning group all of which provided vocational training for the students. Students themselves had keys for the buildings and were responsible for all that went on in the school. Rogozin gained mention in the Israeli press as the school 'best loved by its students' and the Ministry of Education chose it as 'The Dream School'. In echo of the global trend, the violence that once characterised this school had vanished (Hecht no date). That this educational movement is blossoming in the world's epicentre of entrenched conflict is surely cause for hope.

For education to produce a peaceful, just and sustainable global village, it needs to do two things: (1) provide a forum where children can explore and express all their emotions freely and responsibly, and thus learn to master the language of human connectedness – and (2) in the style of the 'emergent curriculum' (aka enquiry-based learning, democratic education, etc.), children's individual loves and passions must be engaged as the engine of learning. It is not enough to say that this kind of education revolution will help create more sustainable and peaceful societies. Developmental brain science would insist that unless schools evolve, the human capacity for large-scale peace and sustainability will remain out of our reach.

The authoritarian, top-down and coercive classroom approaches that we all remember are destined to go the way of the dinosaur. The new educational model is the death knell to competition-based and authoritarian traditions. Though it comes in a broad spectrum of radical to conservative styles, bearing many names and brands and diverse jargons, a core theme runs through all adherents: that of treasuring every child's individual passions and learning styles. I believe the day for dubbing this as 'alternative education' has passed – this is the educational imperative for the future of the human family.

The most important questions that all children should be asked, regularly and often, from their first day at preschool to their last day in high school, are: 'how do you feel?', 'what do you think?' and 'what do you love?' A world of peace, justice and sustainability is entirely attainable; it requires us to begin by relating to children peacefully, justly and sustainably.

Notes

1 Pew Research Center, *Global Public Opinion in the Bush Years (2001–2008)*. A poll of over 175,000 people in 54 nations, media release 18 December 2008 http://pewresearch.org/pubs/1059/global-opinion-bush-years, accessed 20 November 2010.

2 http://www.endcorporalpunishment.org/pages/frame.html, accessed 20 November 2010.

3 http://www.caica.org, accessed 20 November 2010, http://www.cafety.org/press, accessed 20 November 2010; http://www.nospank.net/boot.htm, accessed 20 November 2010.

4 http://childrightscampaign.org/crcindex.php , accessed 20 November 2010; http://www.cwla.org/advocacy/crc.htm, accessed 20 November 2010.

5 United Nations Children's Fund (2007) *Child Poverty in Perspective: An Overview of Child Wellbeing in Rich Countries*, February, http://devon.freepgs.com/2007/UN-child-wellbeing-report.php, accessed 20 November 2010.

6 Ibid.

7 http://en.wikipedia.org/wiki/Making_Sweden_an_Oil-Free_Society, accessed 20 November 2010.

8 http://www.kinma.nsw.edu.au/, accessed 20 November 2010.

9 http://www.transformingconflict.org/index.html, accessed 20 November 2010.

10 Ibid.

11 http://www.democratic.co.il/en/, accessed 20 November 2010.

12 http://www.idenetwork.org/idec-newsletters/idec-newsletters-english.htm, accessed 20 November 2010; http://en.wikipedia.org/wiki/International_Democratic_Education_Conference, accessed 20 November 2010.

References

Cooper, G., Hoffman, K., Powell, B. and Marvin, R. (2005) 'The circle of security intervention: Differential diagnosis and differential treatment', in Berlin, L., Ziiv, Y., Amaya-Jackson, L. and Greenberg, M. (eds) *Enhancing Early Attachments*, New York: The Guilford Press.

Dogakinai, A. (2010) *Ijime: A Social Illness of Japan*, http://www.humiliationstudies.org/news-old/archives/000120.html, accessed 20 November 2010.

Dolby, R., Swan. B. and Croll, J. (2004) *Developing the Community Within a Preschool Setting: How Strengthening the Relationships Between Staff, Children, and Families Achieves Positive Outcomes for the Children*, paper presented at WAIMH 9th World Congress, Melbourne, Australia, 17 January.

Faiola, A. (2004) 'Youth violence has Japan struggling for answers: 11-year-old's killing of classmate puts spotlight on sudden acts of rage', *Washington Post Foreign Service*, pA01, 9 August, http://www.washingtonpost.com/wp-dyn/articles/A50678-2004Aug8.html, accessed 20 November 2010.

Fredman, L. (1995) 'Bullied to death in Japan (teenagers' suicides)', *World Press Review,* 1 March, 25.

Gerhardt, S. (2004) *Why Love Matters – How Affection Shapes a Baby's Brain*, Hove and New York: Brunner-Routledge.

Goleman, D (2006) *Social Intelligence – the New Science of Human Relationships*, London: Hutchinson.

Gordenker, A. (2002) 'School refusers: When one-size-fits-all schooling doesn't fit', *The Japan Times Online*, 3 May, http://search.japantimes.co.jp/cgi-bin/ek20020503ag.html, accessed 20 November 2010.

Grille, R. (2005) *Parenting for a Peaceful World*, Sydney: Longueville Media.

Hannam, D. (2001) *A Pilot Study to Evaluate the Impact of the Student Participation Aspects of the Citizenship Order on Standards of Education in Secondary Schools*, report for Department for Education and Employment, April http://www.csveducation.org.uk/downloads/research-and-reports/Impact-of-Citizenship-Education-Report.pdf, accessed 20 November 2010.

Hecht, Y. (no date) *Democratic Education – A Beginning of a Story*, unpublished manuscript, Chapter 6, Democratic Education Institute, Tel Aviv.

Jones, E. and Nimmo, J. (1994) *Emergent Curriculum*, Washington DC: NAEYC.

Lewis, T., Amini, F. and Lannon, R. (2000) *A General Theory of Love*, New York: Vintage Books.

Lillard, A. and Else-Quest, N. (2006) 'Evaluating Montessori education: An analysis of students' academic and social scores compares a Montessori school with other elementary school education programs', *Science* 313, 29 September, 1893.

Liptak, A. (2010) 'Justices, 5-4, reject corporate spending limit', *The New York Times*, 22 January, pA1, http://www.nytimes.com/2010/01/22/us/politics/22scotus.html, accessed 20 November 2010.

Milburn, M.A. and Conrad, M.A. (1996) *The Politics of Denial*, Cambridge: MIT Press.

Mishel, L. (2006) *CEO Minimum-Wage Ratio Soars*, Washington, DC: Economic Policy Institute, http://www.epi.org/economic_snapshots/entry/webfeatures_snapshots_20060627/, accessed 20 November 2010.

O'Connor, F. (2010) 'Expanding the potential and possibilities of the early years of primary school: A system's challenge', *Curriculum Leadership*, <http://cmslive.curriculum.edu.au/leader/default.asp?id=30320&issueID=12063, accessed 20 November 2010.

Perry, B.D. (1997) 'Incubated in terror: Neurodevelopmental factors in the cycle of violence', in Osofsky, J.D. (ed.) *Children in a Violent Society*, New York: The Guilford Press.

Prosser, T. (2009) *Executive Compensation and the Economic Crisis, Negotiating out of the Crisis*, 25–27 November, http://forum2009.itcilo.org/en/thematical-papers/executive-compensation-and-the-economic-crisis, accessed 20 November 2010.

Schore, A. (1994) *Affect Regulation and the Origin of the Self – the Neurobiology of Emotional Development*, Hillsdale, NJ: Lawrence Erlbaum Associates.

Siegel, D.J. (1996) *The Developing Mind, How Relationships and the Brain Interact to Shape Who We Are*, New York: The Guilford Press.

Soling, C. (2009) *The War on Kids*, Spectacle Films, http://www.thewaronkids.com/MAIN.html, accessed 20 November 2010.

Further reading

Jones, E., Evans, K. and Stritzel, K. (2001) *The Lively Kindergarten: Emergent Curriculum in Action*, Washington DC: NAEYC.

Kakuchi, S. (2003) 'Alternative schools offer hope for "drop-outs"', *Japan Inter Press Service*, 4 June, http://www.asianresearch.org/articles/1272.html, accessed 20 November 2010.

Lakoff, G. (2004) *Don't Think of an Elephant*, Vermont: Chelsea Green Publishing.

18

ECOLOGICAL UNDERSTANDING AND DRAMA

David Wright

> *The pattern which connects:* Why do schools teach almost nothing of the pattern which connects? . . . What's wrong with them? What pattern connects the crab to the lobster and the orchid to the primrose and all four of them to me? And me to you? And all six of us to the amoeba in one direction and to the back-ward schizophrenic in another?
>
> (Bateson, 1979: 8)

Introduction

Drama processes construct more than drama. They construct relationships. These are generally thought of as person-to-person relationships but they can also be relationships to groups or communities: inmates of a nursing home, school students, 'strangers on a train'; relationships to sensed experiences: sounds, feelings, images, emotions; relationships to places, spaces or settings: a windswept beach, a deserted railway station, an inner-city apartment; relationships to other than human life: broccoli, mosquitoes, puppies, poltergeists; and of course, relationships to social and personal processes: racism, celebration, imagination, grief, dreams, ideas and systems of belief or understanding. These relationships do not stand in isolation. They are interwoven: a place, a person, a sound and a feeling are connected in consciousness. Networks of associations arise. Layers of association arise also through questions around qualities of relationship: depth, strength, effectiveness, responsiveness, sustainability and more can be subject matter here.

The qualities and the interlinking of relationships are also the basis of ecology and ecological understanding (Bateson 1972, 1979). Generally, ecology is thought to refer to studies of the relations of organisms to one another and their surroundings (Moore 2007). By contrast ecological understanding, as described by Bateson, applies this understanding, while admitting self-reflective humans to the process. Importantly, understandings of relationships that include humans are not pre-formed. These relationships are dependent on culture. A city dweller encounters the world differently to a farmer, a Christian differently to a pagan. 'What we observe', wrote Heisenberg, 'is not nature, but nature exposed to our questioning' (cited in Capra 1996: 40). Like Boal's 'spect-actors' (1992) none of us can stand apart from that which we are participating in.

Aims

The aim of this chapter is to look at understandings enabled through drama and to discuss those understandings in relation to a discourse around ecological understanding. The intention is to open space for appreciation of drama as a critically reflective methodology suited to the development of an ecological way of thinking, then to consider the broader relevance of insights arrived at in this way. In progressing the discussion I want to draw on the concept of ecological understanding through reference to the work of Gregory Bateson (1972, 1979; Harries-Jones 1995).

Bateson and ecology

Bateson was a biologist and as such interested in relationships and patterns in nature. But he saw his work as more than this. Bateson was interested in the ways of thinking that enable relationships and patterns in nature to be identified and appreciated. This is a significant extension. Bateson argued that the orthodox scientific perspective, which views ecosystems as entities that can be observed, controlled and managed, is flawed. Instead he offered an ecological epistemology based on the premise that our survival depends on understanding that we are intimately coupled to our vision of the ecological order. This, Bateson argued, determines how we think and act in relation to that order. Thus our appreciation of an ecosystem is limited by our imagination of that which is included in its field. Limits to understanding construct limits to action, hence the significance of questions around the degree to which humans are part of the system. This suggests that identifying ecology as an issue of politics or commerce trivialises a deep and far-reaching study and almost inevitably compromises action. The action Bateson sought, and it is worth remembering he offered these thoughts long before ecology became front-page news, is action that will 'unify' and 'sanctify' the natural world (Bateson 1979: 19).

Bateson's unity was not based on an anticipation of scientific discovery or courageous political decision making. It was based on an assumption that we humans have the capacity to act upon our appreciation of interrelatedness and will, as a consequence, arrive at a practical recognition that we cannot totally control the natural systems of which we are a part. These assumptions led him to his interest in the role of learning in transformation. It also led him to his most potent metaphor: addiction.

Bateson's learning about addiction arose in his studies of Alcoholics Anonymous (AA). Central within AA programs is the recognition that addiction cannot be controlled. 'The panic of the alcoholic who has hit rock bottom is the panic of the man [sic] who thought he had control over a vehicle but suddenly finds that the vehicle can runaway with him' (cited in Harries-Jones 1995: 42). The AA cure requires a new way of thinking through reference to 'a greater power than myself'. This is an example of the sort of 'higher order learning', which requires individuals to arrive at and live through new assumptions. It is the sort of learning that Bateson believes is needed to overcome ecological problems.

> People fail to realize that they are addicted to patterns of consumption that disrupt a normal healthy relationship between themselves and the environment of which they are a part. A 'cure' to this addiction is required through reorganization of Western society's relations with its environment. No reordering of ecological context is possible until people realize that the source of environmental stress is their false sense of control and the patterns of fixation they have formed in their relationship with the environmental systems of which they are a part.
>
> (Harries-Jones 1995: 56)

Systems theory

Systems theorists like Bateson argue that there is a formal relationship between natural systems and cognition. This analysis is extended in self-organising systems theory, which works with the notion that cognition or mind is an integral part of the system it seeks to explain. Systemic approaches have influenced work in the sciences, particularly biology, where organic life has been analysed through reference to its self-organisation (Maturana and Varela 1987; Mingers 1995; Capra 1996). This suggests that living organisms have the capacity to organise aspects of their own existence in response to environmental influences. Some theorists, most notably Niklas Luhmann (1995), have inferred the social from the biological and extended this theorising into the realms of communication.

Luhmann argued that 'humans cannot communicate; not even their brains can communicate; not even their conscious minds can communicate. Only communication can communicate' (Luhmann cited in Moeller 2006: 8). By focusing on the relationships required for communication Luhmann asserts that society is a self-organising network of communication systems. These systems comprise and organise those who participate in them. These systems include, for example, business, the political system, the legal system, the education system, the arts and more. Each system is culturally positioned and linguistically defined. Each is built upon a history of attitudes and practices. Each contains and focuses the work of those who identify with and participate in them, and each changes within shifting boundaries over time.

Drama is also a communication system: one that works through a meta-analysis of communication processes. Interrelationship is always present and always subject matter in drama. In this respect drama is a laboratory for the conscious exploration of relationships. Accordingly, one of the principal functions of an education in drama is the facilitation of a conscious awareness of how communication unfolds and the circumstances that influence that unfolding. Alongside this, drama can provide an education in the capacity to employ that conscious awareness. This occurs in drama and arises also as a consequence in life. In this respect epistemology and ontology are combined in the generation of practice. Knowing is doing. This systemic awareness is far-reaching. It affirms the way in which drama constructs more than drama. Drama constructs relationships or, more accurately, drama constructs an enriched awareness of the relationships we are participating in.

Learning in drama

The bedrock of drama is trust. O'Toole argues that teaching drama 'demands an understanding of how to frame a situation so as to provide protection and permission, as well as dramatic tension and action' (O'Toole in Sinclair et al. 2009: 10). In the same publication Reid writes; 'As a teacher of the arts, there is no sweeter moment for me than when the shy child steps to the front of the class and accepts the creative risk (Sinclair et al. 2009: 55). While it is the responsibility of the teacher to provide 'protection and permission' to enable the 'shy child' to step forward it is the trust or learning of the child that enables that step. Students need to feel confident their imagined worlds will not be undermined. Drama games are often used for this purpose and children are invited to play with the elements of drama through these. This is where the origins of drama education lie: children's play (Slade 1976; Hornbrook 1989; Burton 1991).

In play an imagined life is engaged. An imagined life is one in which relationships are transformed: a cardboard box becomes an aeroplane, a rock becomes a mountain, a friend becomes a policeman and I become a superhero. Trust enables transformation to be sustained. When sustained, transformations determine further relationships and the environments they work through: the pilot in the aeroplane in the clouds in the sky, the policeman in the park who encounters good guys and bad guys, the superhero with a secret identity as a chef who cooks pizza and chocolate pies.

More recently, as secondary school drama has drawn on actor training, play has been formalised as 'improvisation'. As well as providing permission to play, improvisation provides a structure for the appreciation of play. Johnstone (1989) and Boal (1992) have made these structures accessible, Johnstone through theatre sports and Boal through 'theatre of the oppressed'. Pierse (1993: 91), who applies Johnstone's work, describes improvisation as a six-stage process.[1] These six can be further reduced to a basic three: 'Offer', an offer is made; 'Accept', the offer is accepted and 'Extend', further offers are made as a result of that which has been accepted. Necessarily, any breakdown between stages halts the process. Extended play and improvisation require both permission and sustainable relationships. These are best constructed with an appreciation of the dynamics that facilitate their accomplishment.

'Playbuilding' (Bray 1991; Lovesy 2003; Hatton and Lovesy 2008; Norris 2009) is a term used to describe a means whereby people come together to devise and perform their own dramatic works. It is also a way of generating and applying knowledge. Like its close relative performance ethnography (Denzin 2003), it is also a means of inquiry through participation in a collective creative enterprise. In that respect its learning outcomes include, but extend beyond, performance. Describing playbuilding as a form of research Norris writes,

> Trust is vital in any process of co-creation, and, since all participants are stakeholders, a respect for one another's position is vital. As director/researcher, one of my duties is to set the tone of our work. I assert that we are all in a state of becoming (de Chardin 1969), and we, as pilgrims, have gathered to tell our stories (Kopp 1972) not only to advance our current positions but also to change them, when personally deemed appropriate, as we listen to the stories of others.
> (Norris 2009: 23)

The 'state of becoming' Norris refers to can be found throughout dramatic experience. Knowledge unfolds as relationships are seen to develop. This occurs in the experience of audience as well as performers. It occurs in drama education particularly because, for the purpose of learning, there is a strong responsibility for performers to also be audience: to have an awareness of their participation in the drama.

Becoming of the sort Norris refers to emerges in relation to numerous factors. These include other participants in the drama, the setting, the subject matter, form, style or narrative. Technologies can be a factor, as can sites and situations along with assumptions about relationships to audience. All of these contribute to an appreciation of the ecology – the network of interrelationships – that comprise the drama. This appreciation is enhanced in extended practice or rehearsal, which involves a tightening focus on relationship construction. Rehearsal prioritises responsiveness to character and environment and an enhanced awareness of personal participation in the system of understanding that is the drama. This ecological understanding, as Bateson observed, arises within an experience of immersion. Depth of involvement limits detached awareness. Participants interact in response to the internal workings of the drama. This defines it as self-organising. An appreciation of this aspect of the drama enables it to be understood as a means of facilitating broad social–ecological learning. Understanding of this kind is applied, necessarily. Thus learning constructs ontological consequences.

Like Norris, Maturana, who along with Varela has had considerable impact upon ecological thinking, also works with the idea of 'becoming'. He uses it to suggest the activity contained in the passive notion of 'being' (1987). Thus Maturana and Varela (1987) observe that – and this notion should be familiar to all drama practitioners – we 'bring forth a world' through our 'manner of living'. O'Sullivan and Taylor (2004: 21) use this perspective to argue that,

> Knowledge in an ecological perspective is firmly imbedded in the practical world ... We generate knowledge from the practice of living and working

– action, reflection, commonsense-making, social construction. That is, knowledge is not individually derived and held but rather generated in relationships with others.

Drama is ideally situated for the exploration of this way of thinking. This is different to the use of drama to argue environmental issues or dramatic performances situated in specific environments or locales. It is an argument for the recognition of drama as a means of facilitating ecological knowing – of the sort alluded to by Bateson, O'Sullivan and Taylor. As a prerequisite, such facilitation requires recognition of this ecological perspective, in the context of contemporary concerns about our collective future.

Drama and ecological understanding

Some of the most influential educators in ecological understanding work with drama processes. Joanna Macy (1991, 1998), whose work centres on 'despair and empowerment' uses ritual, dance, emotional identification, mask, costume and other drama tools and activities to 'explore the inner resources that are needed to take action in today's world' (1991: 3). In keeping with Macy's Buddhist orientation many of these activities have a meditative character, while others have a powerful revelatory intent, most particularly the 'Council of All Beings' (Seed et al. 1988). This is a flexible workshop activity devised by Macy and Seed, which can last several days. It uses drama processes such as improvisation, naming games, milling, mask-work, storytelling, character work, emotional identification and structured presentation and reflection for the purpose of ritual identification with the natural environment. While its focus is environmental it is the deep emotional processes of interrelationship and identification that mark it as a prompt to ecological knowing.

In my own work with students I regularly use drama activities. In a recent undergraduate drama method class we looked at Heathcote's 'mantle of the expert' exercise. This exercise imagines life in the shadow of a medieval monastery (Heathcote and Bolton 1994), as a means of interrogating social power. It soon became apparent that none of my Australian students lived in a village overlooked by a monastery. We talked about that which did oversee our homes. One student, who lives in the nearby Blue Mountains,[2] recognised that her home was overshadowed, especially each summer, by the possibility of bushfire. This became our subject matter and our play-building then focused on our collective relationship to bushfire: its destructive power accompanied by its power in building community along with its capacity to stimulate regeneration, regrowth and new life. During an intense three-hour class we gathered tales of powerful visceral encounters with fire. These told of the fear of fire, the piercing noise of warning sirens, the domination of the media by safety warnings, the red fire-smoke clouds that become the sky, the charred fragrance of burning eucalyptus followed some time later by the return of bird life, the recovery of reptiles, the sweet smell of new leaves sprouting fresh from charcoaled bark. Students encountered a communal learning about fire, but more than this, their learning included a deepened understanding of how peers meet powerful existential challenge. It became an exercise in the building of community through vulnerability and trust. Drama was the medium of the learning and fire the immediate subject matter but interrelationship was the substance of the class. 'I learned about fire as a collective experience', said Denise.

However, ecological understanding relates to more than the appreciation of environmental problems. The story above describes a relationship between fire and consciousness. While the fire is important it is the relationship that determines the learning. This sort of learning was the focus of a 2008 workshop led by adult learning theorist Peter Willis. Willis described the workshop as 're-enchantment education', employing social–ecological ways of thinking. The process entailed, Willis said, 'an exploration of thc stories that give meaning to life' (Willis personal communication 2009).

Willis asked participants to take on characters contained in an extract from a film he showed. The film, *As it is in Heaven*, depicts conflicts within a Swedish village community as a choir forms, rehearses and performs, under the direction of a disenchanted musical prodigy. By taking on roles and encountering conflicts experienced by characters in the community, Willis sought to inculcate 'a phenomenological encounter with the construction of relationships in learning' (op cit). Willis asked, how do these people learn? Then, how do I learn under these circumstances?

The learning that unfolded was bound within two environments, the fictional environment circumscribed by the film and the real world workshop environment within which the film was re-enacted. Thus both fictional self and workshop self became characters available for analysis. Their learning became subject matter. Willis argues that such exercises offer an opportunity for an educator to 'become aware of how … [teaching and learning] sits with [personal] aspirations and self stories' (Willis personal communication 2009). These aspirations and self-stories are directly related to the relationships constructed in the workshop. The ecology of the workshop therefore both serves and emerges as a consequence of learning generated therein.

All social interaction can be discussed in dramatic terms. Schechner's (1977) analysis of performance, which draws on Turner (1982), contributes to an appreciation of the boundaries within such interaction occurs. The insight made available through the vocabulary of drama: such terms as 'context', 'setting', 'character', 'role', 'relationship', 'focus', 'tension', 'movement', 'symbols', 'mood', 'drama' itself and more is both succinct and extremely helpful, in this regard. It enables processes to be tracked and articulated and is particularly useful in the observation of learning. Given the relationship between drama and ecological understanding, social interaction, and the participants therein, can also be interpreted through the lens of ecology: through reference to the relationships that construct learning. The notion of 'learning ecology' (Hill et al. 2004: 49–50) highlights this.

> Learning ecology provides a means for understanding and working with the complex and diverse ways in which individuals … learn, become more conscious, develop worldviews, change and act on their values. It takes a constructivist view and acknowledges how our previous life experiences and opportunities, interactions, learning styles, and personalities result in each individual having a unique learning ecology.

Concluding comments

The use of drama to heighten ecological understanding incorporates and extends Nicholson's focus on applied drama as the ethical construction of a 'more generous and multi-faceted world by making a creative space in which fixed and inequitable reality, identity and difference might be disrupted and challenged' (2005: 167). Here the encounter is understood in systemic terms. Ecology is introduced in terms of both process and subject matter, self and context are examined and meaning is located in a heightened appreciation of deep, systemic interrelationships. The 'other than human' world is incorporated wholeheartedly and the learning contained therein enriched accordingly.

Work of this kind promotes self-reflection in the broad context of considerations on our place in the complex web of emerging systems, patterns and relationships. This is a radical approach to education. It is in accord with O'Sullivan's (1999: xiv) argument that,

> Every profession and occupation of humans must establish itself within the integral functioning of the planet. The earth is the primary teacher in economics, in medicine, in law, in religion. Earth is the primary educator.

An early step in any such appreciation is an invitation to acknowledge our participation in the systems and relationships that are being studied. As David Orr reminds us, the fact that many things on which our future health and prosperity depend are in dire jeopardy is not the product of ignorant people. It is the consequence of work by highly educated people (Orr 2004). The issue is not therefore one of education or learning, it is the assumptions upon which that education or learning is constructed.

For those working with drama, ecological understanding does not simply mean drama about ecology or ecological systems or ecological problems. This sort of drama/theatre is valid and of interest but different in intent. It has been canvassed widely (for example Coult and Kershaw 1990; Marranca 1996; Kershaw 2007). The ecological approach that is being alluded to here enables the realisation that drama, which is a study of relationships, learning and systems of communication, facilitates ecological insight, which facilitates, in turn, ecological literacy: Bateson's 'pattern that connects'. This realisation constructs a deepened appreciation of the social dimensions of ecological systems and ecological understanding as a consequence of the realisation that ecology, society and epistemology are inextricable. The key elements of drama are at play in these social dimensions. Necessarily, an education process that draws attention to these relationships also, when successful, brings us together in social understanding. It initiates mutuality and belonging. Beyond the yearning for belonging, which attracts so many students to drama, it looks towards the deeply connected and respectful relationships that are the basis of sustainable communities and resilient futures.

Notes

1 Pierse's (1993) six stage process: Make an offer; Yield to the offer; Name the offer; Extend the offer; Advance the offer; Shelve an offer.

2 These 'Blue Mountains' rise in a vast world heritage area, to the west of Sydney, Australia. It separates the coastal fringe from the western plains. It has a population of about 70,000, most of whom live in a string of towns that follow the rail line through a vast, undulating, rocky, forested wilderness.

References

Bateson, Gregory (1972) *Steps to an Ecology of Mind*, New York: Ballantine Books.

Bateson, Gregory (1979) *Mind and Nature*, New York: Bantam Books.

Boal, Augusto (1992) *Games for actors and Non-actors*, London: Routledge.

Bray, Errol (1991) *Playbuilding*, Sydney: Currency Press.

Burton, Bruce (1991) *The act of Learning*, Melbourne: Longman Cheshire.

Capra, Fritjof (1996) *The web of life*, London: Harper Collins.

Coult, Tony and Kershaw, Baz (eds) (1990) *Engineers of the Imagination: The Welfare State Handbook*, London: Methuen.

De Chardin, Pierre Teilhard (1969) *Christianity and Evolution*, New York: Harcourt, Brace, Jovanovich.

Denzin, Norman K. (2003) *Performance Ethnography*, Thousand Oaks, CA: Sage.

Harries-Jones, Peter (1995) *A Recursive Vision: Ecological Understanding and Gregory Bateson*, Toronto: University of Toronto Press.

Hatton, Christine and Lovesy, Sarah (2008) *Young at Art: Classroom Playbuilding in Practice*, New York: Routledge.

Heathcote, Dorothy and Bolton, Gavin (1994) *Drama for Learning*, Portsmouth, NH: Heinemann.

Hill, Stuart, B., Wilson, Steve and Watson, Kevin (2004) 'Learning ecology. A new approach to learning and transforming ecological consciousness', in O'Sullivan, Edmund and Taylor, Marilyn (eds) *Learning Towards an Ecological Consciousness*, New York: Palgrave Macmillan.

Hornbrook, David (1989) *Education and Dramatic Art*, Oxford: Blackwell Education.

Johnstone, Keith (1989) *Impro*, London: Methuen.

Kershaw, Baz (2007) *Theatre Ecology*, Cambridge: Cambridge University Press.

Kopp, Sheldon (1972) *If you Meet the Buddha on the Road, Kill Him*, New York: Bantam.

Lovesy, Sarah (2003) *Drama Education Secondary School Playbuilding*, unpublished PhD thesis, http://handle.uws.edu.au:8081/1959.7/22408, accessed 15 April 2010.

Luhmann, Niklas (1995) *Social Systems*, Stanford, CA: Stanford University Press.

Macy, Joanna (1991) *World as Lover, World as Self*, Berkeley, CA: Parallax Press.

Macy, Joanna (1998) *Coming Back to Life*, Stony Creek, CT: New Society Publishers.
Marranca, Bonnie (1996) *Ecologies of Theatre*, Baltimore, MD: Johns Hopkins University Press.
Maturana, Humberto and Varela, Francisco (1987) *The Tree of Knowledge*, Boston, MA: Shambhala.
Mingers, John (1995) *Self-producing Systems*, New York: Plenum Press.
Moeller, Hans-Georg (2006) *Luhmann Explained*, Chicago, IL: Open Court.
Moore, E. (ed.) (2007) *Australian Pocket Oxford Dictionary*, South Melbourne, Victoria: Oxford University Press.
Nicholson, Helen (2005) *Applied Drama*, New York: Palgrave Macmillan.
Norris, Joe (2009) *Playbuilding as Qualitative Research*, Walnut Creek, CA: Left Coast Press.
O'Sullivan, Edmund (1999) *Transformative Learning*, London: Zed Books.
O'Sullivan, Edmund and Taylor, Marilyn (2004) *Learning Towards an Ecological Consciousness*, New York: Palgrave Macmillan.
Orr, David (2004) *Earth in Mind*, Washington, DC: Island Press.
Pierse, Lyn (1993) *Theatresports Downunder*, Sydney: Improcorp Australia.
Schechner, Richard (1977) *Essays on Performance Theory 1970–1976*, New York: Drama Book Specialists.
Seed, John, Fleming, Pat, Macy, Joanna and Naess, Arne (1988) *Thinking Like a Mountain: Towards a Council of All Beings*, Philadelphia, PA: New Society Publishers.
Sinclair, Christine, Jeanneret, Neryl and O'Toole, John (eds) (2009) *Education in the Arts*, South Melbourne, Victoria: Oxford University Press.
Slade, Peter (1976) *Child Drama*, London: Hodder & Stoughton.
Turner, Victor (1982) *From Ritual to Theatre*, New York. PAJ Publications.

19

DRAMATIC PLAYFULNESS AND THE HUMAN SPIRIT

Graeme Frauenfelder

> Don't ask what the world needs.
> Ask what makes you come alive, and go do it.
> Because what the world needs is people who have come alive.
> (Howard Thurman)

I am fascinated by how the arts and creative playfulness enhance effectiveness and enrich human spirit in international aid and community development projects. I was inspired by Barbara, a Zambian PhD student, when she spoke to a social ecology university class about how her people needed renewing and rebuilding of their spirit and culture as much as they needed food, water, health care, education and economic aid. Aliveness of spirit gives motivation and empowerment to live, heal, develop and create new visions of possibility.

From my first days of fieldwork research in Zambia I dramatically saw the impact of cultural and creative expression. Initially it was demonstrated through Enias Jere, the community empowerment specialist I worked with. From his drab western-style office attire in the city he transformed into an African performer in colourful, artistic Nigerian costume in the villages. In contrast to hardly speaking in office staff meetings, he became a charismatic and engaging facilitator in Nega Nega and Chadiza villages. His eyes sparkled and he radiated aliveness as he danced, sang, acted, told stories and recited poetry. The effect was similar on his workshop participants. His colleague Gina Chiwela was curious to understand the transformation. Because of this, Enias was hired to integrate cultural creativity into the village-based projects of People's Action Forum (PAF: a Zambian community empowerment organisation).

This phenomenon of 'aliveness' through active engagement in creative activities lies at the heart of Cultural Community Development, which has several underlying values:

1 Empowerment of individuals and communities at the grassroots level through processes that enhance choice, and extend connection and sense of community, which also assist them with envisioning and creating their future (Adams and Goldbard 2002).

2 Valuing cultural and artistic activities to enrich personal and community life, leading to renewal and adaptation in the face of contemporary challenges (Adams and Goldbard 2002).
3 Focusing on social justice and human rights for all, particularly the marginalised (Paranjape 2002), and promoting the highest level of human values (Couratier and Quinn 1998).
4 Commitment to sustainability. A social ecology perspective adds an environmental emphasis to the valuing of cultural and economic sustainability (Schafer 2003).

'Enriching the human spirit' lends itself more to poetic and artistic description than precise definition. It's hard to measure using facts, statistics and graphs that aid donors and institutions demand (Crawford 2004). Yet it is clearly seen in the sparkle in eyes, vitality in bodies, emotional expressiveness, aliveness in relationships and the collective vibrancy of a community, as well as in their visions for the future. Embodied engagement in the creative processes of the arts brings present moment aliveness, while enabling personal, community and planetary transformation (Nachmanovitch 1990). For sustainability, gaining intrinsic satisfaction and fulfilment through creative and cultural pursuits is a healthy contrast to extrinsic, materialistic, consumption-based lifestyles (Shafer 2003).

I invite you on a journey through stories and reflections demonstrating the enlivening and transformative power of the arts. They come from personal experiences of awakening the human spirit through playfulness and performing: as a clown in rural China and urban South Africa, as a community empowerment volunteer in Zambian villages, and from university research.

Awakening the Spirit

The 2008 Beijing Olympic festivities and sporting events captured the attention of the world, while millions of Chinese citizens were traumatised in the aftermath of the Sichuan province earthquake. In displaced people's camps, each family was provided with basic food and a small room in one of many long, temporary buildings. Grief and boredom were common. Parents were worried their children were permanently scarred emotionally as some hadn't laughed or smiled in months. A team of international and Chinese volunteers, coordinated by Fusion, were invited to a camp in the mountains. Fusion is an international, Australian-originated, community and youth work non-profit organisation. As one of the volunteers in China I had the role of being a clown and clown-trainer. We created interactive, playful festivals to generate community spirit and provide enrichment to people who had lost almost everything.

Among other benefits, creative activities distract people's minds from problems and provide a peaceful break in stressful times (Heritage 2002). Rather than it being premature to focus on culture when there is material deprivation and struggle, Amartya Sen (1998: 317) insists: 'Culture does not exist independently of material concerns, nor does it stand patiently waiting its turn behind them.'

Four nights of interactive festivals drew people into playful, creative activities. Group games, dancing, skipping, hula hoops, stilt walking, juggling, face painting, paper folding, paper aeroplane competitions and other physical and artistic activities initially attracted children but soon all ages joined in. Children and dads helped grandpas walk on stilts. Grandmas line-danced with teenagers. Mums skipped with kids. Strangers played juggling games together after making their own juggling balls with rice and balloons. Teenage girls became face-painters while teenage boys teased and mimicked clowns. A few soldiers, the camp security guards, forgot their roles, let down their guard and danced and played free-spiritedly. Through creating playful interaction and showing that the outside world remembered and cared, the enthusiasm and community spirit grew each evening.

By the last night, many of the activities were led by people from the displaced people's camp as they embraced and embodied the festive spirit. This spirit went beyond fun: there was also a deeper sense of love, care and respect. Some of the festival team continued to visit and support the community. The experiences and stories from this experience are numerous. They demonstrate that arts definitely have a role to play in enhancing quality of life and 'lifting the human spirit' (Ybarro-Fausto 2002: 6), especially in times of despair, and they have a therapeutic role in bringing healing and support (Legarda 2002).

Creating and deepening human connections

A summer Sunday in South Africa holds powerful memories for me. On a day off from preparing for community festivals I played with lion cubs, hugged giraffes and explored an archaeological site considered to be the birthplace of some of the earliest humans – then faced death at gunpoint in a Johannesburg park. In the after-reflections I imagined the violent men as once being young, playful and open-hearted like the children who walked the dogs down the street with us each day, helped make festival equipment and led me, the clown, hand in hand to all the festival activities in Alexandra township. My heart ached at what had brought these men to this point, which soon led to jail.

The community-building projects took on new personal meaning. They were part of repairing social fabric that was fragile at times, and included reclaiming public space for community interaction. Connections and rapport were built with people across a diversity of backgrounds, generating much needed social capital (Putnam et al. 2003).

The next festival – at Brixton – was two days after the hold-up. The site was strategically positioned at the intersection of several distinct communities of varying socio-economic and cultural backgrounds. Interactive activities surrounded the central games area and people were encouraged to join in the large group activities. Barriers of colour, culture, economics, gender and language were transcended as adults and children laughed, danced, played, sang and held hands. Cultural activities like this prove to be beneficial as they 'work effectively with an enormous range of social and age groups' (Adams and Goldbard 2002: 28) and strengthen people who face oppressive forces and trends (Legarda 2002), such as the violence and social fragmentation in Johannesburg.

Treasured memories include a burqua-clad Muslim woman making a giraffe out of balloons for two Afrikaans-speaking boys, while a Zulu-background clown playfully interacts with them. In the heart of the activities is a dreadlocked Rastafarian white guy, slightly spaced out, asking to go on the road with us to get more of the vibe. We were begged to do this again, showing enthusiasm for continuing experiences of the playful spirit and connection across social barriers.

Expanding awareness and deepening empathy

A crowded bus took Enias and I far from Lusaka, the Zambian capital, until it broke down and left us stranded in the scorching summer sun. Eventually we found a car ride to Nega Nega village on a rough yet scenic drive through sugarcane plantations, past mud and thatch huts and finally to the government school. About twenty teenagers and several adults were training as an anti AIDS drama group. They planned to educate their community about the sensitive and emotionally-laden issues surrounding HIV/AIDS, with which 20 per cent of Zambians were considered to be infected.

The group performed several short skits demonstrating their current understanding, focusing on behavioural directives through simplistic slogans including variations of 'Use condoms' and 'Don't have sex'. One leader was judgemental regarding people with HIV/AIDS and dogmatic about his beliefs about dealing with HIV/AIDS. We quickly realised the participants needed to deepen their awareness and develop empathy before creating educational performances. These were enhanced by expanding imagination and exploring personal and cultural stories, which the arts do well (Arnold 2005).

Using drama activities Enias took the participants on an insightful inner journey. They gained new perspectives playing roles other than their own. Judgement shifted to empathy. Simplistic behavioural slogans gave way to deeply exploring the issues and considering ways to empower people affected by HIV/AIDS. Songs were used to reinforce learning and became part of the final performance.

The young men created a skit demonstrating dynamics that drew them and their peers into unsafe sex. The young women devised one about social pressures and cultural practices that threatened them. The adults' drama explored marriage and extramarital factors relating to HIV/AIDS, family dynamics and intimate relationships. All three were interwoven to create a street theatre performance about a family challenged and torn apart by HIV/AIDS. Characters represented different generations (children, youth, adults and elders), female and male genders, single and married people, and people infected and not infected with HIV. Everyone in the audience would have characters to identify with.

The plot was true to life and emotionally engaging as it explored teenage sexuality, parenting issues, extramarital sexual activity, marital conflict, family love, and facing social stigma and death from HIV/AIDS. It was insightful and motivating without being moralistic or judgemental, and challenged taboos around discussing sexuality and cultural issues to do with HIV/AIDS. The man who was judgemental on the workshop's first day, moved many of us to tears as he took on the character of a lonely, dying father, showing significant personal transformation as he arrived at deep empathy and understanding of the complexity and humanness of the issues.

The finale for the workshop was dynamic. A couple of hundred people gathered in the village marketplace to experience the show. This was significant entertainment as performances weren't common and the village had no TVs, movies, clubs or theatres. One of the benefits of educating through performance is that it draws and holds the interest of large numbers of people (Adams and Goldbard 2002). The audience was drawn into the dancing that concluded the show. The drama was performed over the following months throughout neighbouring villages and in schools, as well as for the entire village on World AIDS day.

A week later, volunteer community facilitators from the Nega Nega area participated in HIV/AIDS awareness training. Drama, art, poetry, dancing and singing were integral to the learning. Most participants were shy or silent in discussing issues around sexuality, yet their role in the community was to break through this silence. In small groups the participants explored the issues through drama, exploring different roles.

Discussion flowed as individuals talked about the story, subplots, characters and dilemmas. We observed how the imaginary and fictitious nature of some forms of the arts, makes it easier to raise difficult and taboo topics that are not often spoken about publicly (Aoun 2002). The arts can also function as a mirror reflecting back what is not always noticed (Oropeza 2002). Freedom to reflect, discuss and go beyond taboos around critical issues as a result of dramatic exploration was a common feature in each workshop I experienced in Zambian villages.

The anti-AIDS drama group from the previous week came by and acted out their street theatre performance to conclude the community educators' training. At the end the workshop participants asked us to wait. They disappeared behind the school and emerged half an hour later with their own show to prove it wasn't just youth who could be creative! We were entertained with several improvised skits followed by poetry and songs, spontaneously reflecting the internalisation of their learning. In the workshop feedback there were many requests for more training in artistic methods of community education, awareness and empowerment as they saw great potential for this type of work after having experienced it with new insight and enjoyment themselves.

In *Pedagogy of the Oppressed*, Paolo Freire (2003) championed what he called 'conscientisation', developing consciousness that creates critical understanding of people's reality in ways that engage them in transformative action (Sleeter et al. 2004). People's Action Forum was inspired by Freire and demonstrated it through their work, as I was privileged to observe. I visited Nega Nega several years later to meet some of the workshop participants and was inspired by the continued empowerment of their communities as a result of their ongoing conscientisation through the arts and other transformational learning experiences.

Creating the future

Community development and cultural change calls for imagining alternatives and creating visions of possibility. Artistic endeavour is 'not just ornamental, an enhancement of life, but a path in itself, a way out of the predictable and conventional' (Poth in Cameron 1996: 87). Safe spaces need to be created for free thought and experimentation as visions are developed and brought to fruition (Stanley 2002).

People's Action Forum works with villagers to empower them to create their own community schools rather than wait passively for hoped-for government schools. Enias, Chola Mtonga (the PAF education consultant) and I facilitated a training week for community schoolteachers and leaders in the Chadiza region of northern Zambia. Most had no formal teaching or leadership training. We used a storytelling and drama approach for the leadership component. I told stories of both effective and ineffective leaders from my experience. The participants shared personal stories in small groups, noting the qualities they valued and the styles that were ineffective or harmful, then developed performances to illustrate their learning. Even though they were reluctant to rehearse, we were delighted with their entertaining and dynamic performances. Included with the acting were stories, songs, poems, dances and re-enactments of cultural rituals.

My intention was for the groups to create fictitious scenarios to illustrate their learning, but it seemed that some were being acted as if they were real life, in real time, and that they weren't going to end until everything was solved! When asked about their stories, most said that they were addressing actual difficulties in their own communities. Using what they had learned and already valued, they were exploring options and possibilities to move on from being stuck and feeling helpless as they had previously been. Intuitively they were doing what Augusto Boal developed as an aspect of his Theatre of the Oppressed: rehearsal for reality (Boal 1992).

Patriarchal abuse of power, misuse of money, lack of consultation, inappropriate teacher–student relationships and cultural challenges were issues experimented with and rehearsed. The people took this opportunity to empower themselves to bring unresolved issues into the open and find ways to create better futures. This demonstrates 'arts activity provid[ing] a sort of a lab or rehearsal for social action' (Adams and Goldbard 2001: 36).

The arts can simply be added to community development projects in ways that enrich the experience, enhance effectiveness and lift the human spirit. While there is a place for this, arts-based cultural community development goes beyond 'what' is included (the arts and culture) to the 'how' and 'why' – these include the underlying principles mentioned earlier: creative cultural enrichment, community empowerment, ethical motivation through social justice and human rights, and holistic sustainability.

Some of my creative work is overtly serious, yet still often playful, like the Zambian village workshops. But in the face of big issues and serious problems in the world, the clowning can seem insignificant, with its crazy costumes, bubble-blowing, hi-5s, hide-and-seek along the streets and dancing with strangers in the centre of a festival. Then I do it again ... and experience the aliveness and awakening of people's spirits. In the centre of the big challenges, there is need for aliveness, for play, for heartfelt connection. So I come alive doing what I love and train others to creatively awaken their own and others' sense of aliveness.

In *Free Play: The Power of Improvisation in Life and the Arts*, Stephen Nachmanovitch (1990: 189) reflects insightfully on playfulness, the arts and social action:

> I would like to say too that we have a right to a beautiful and healthy world. But this is not so; art and a beautiful world are made by hard work and free play ... Our work links art and survival, art and healing, art and social change ... Culture and the arts are a vital resource for survival ... and a primary resource of human realisation.
>
> In my work and broader life experiences, I have experienced time and time again what Howard Thurman (2010) calls for in the world: '... people who are alive.'

References

Adams, D. and Goldbard, A. (2001) *Creative Community: The Art of Culture and Development*, New York: Rockefeller Foundation.

Adams, D. and Goldbard, A. (2002) *Community, Culture and Globalization*, New York: Rockefeller Foundation.

Aoun, I. (2002) 'Different art forms: Mutual concerns', in Adams, D. and Goldbard, A. (eds) *Community, Culture and Globalization*, New York: Rockefeller Foundation.

Arnold, R. (2005) *Empathic Intelligence: Teaching, Learning, Relating*, Sydney: University of New South Wales Press.

Boal, A. (1992) *Games for Actors and Non-Actors*, London: Routledge.

Cameron, J. (1996) *The Vein of Gold: A Journey to Your Creative Heart*, London: Pan Books.

Couratier, M. and Quinn, M. (eds) (1998) *World Culture Report 1998: Culture, Creativity and Markets*, Paris: UNESCO.

Crawford, P. (2004) Interview with Paul Crawford (conducted by Graeme Frauenfelder on 1 November 2004).

Freire, P. (2003) *Pedagogy of the Oppressed*, New York: Continuum.

Heritage, P. (2002) 'Real social ties? The ins and outs of making theatre in prisons', in Adams, D. and Goldbard, A. (eds) *Community, Culture and Globalization*, New York: Rockefeller Foundation.

Legarda, M. (2002) 'Imagined communities: PETA's community, culture and development experience', in Adams, D. and Goldbard, A. (eds) *Community, Culture and Globalization*, New York: Rockefeller Foundation.

Nachmanovitch, S. (1990) *Free Play: The Power of Improvisation in Life and the Arts*, New York: GP Putnam's Sons.

Oropeza, M. (2002) 'Huehuepohualli: Counting the ancestor's heartbeat', in Adams, D. and Goldbard, A. (eds) *Community, Culture and Globalization*, New York: Rockefeller Foundation.

Paranjape, N. (2002) 'Small media big potential!', in Adams, D. and Goldbard, A. (eds) *Community, Culture and Globalization*, New York: Rockefeller Foundation.

Putnam, R., Feldstein, L. and Cohen, D. (2003) *Better Together: Restoring the American Community*, New York: Simon & Schuster.

Schafer, D. (2003) 'Diversity and sustainable development: Contemporary concerns or permanent realities?', in Cvjeticanin, B. (ed.) *Culturelink Review: Special Issue 2002/2003: Cultural Diversity and Sustainable Development*, Zagreb: Institute for International Relations.

Sen, A. (1998) 'Culture, freedom and independence', in Couratier, M. and Quinn, M. (eds) *World Culture Report 1998: Culture, Creativity and Markets*, Paris: UNESCO.

Sleeter, C., Torres, M. and Laughlin, P. (2004) 'Scaffolding conscientization through inquiry in teacher education', in *Teacher Education Quarterly, Winter* 2004, http://www.teqjournal.org/backvols/2004/31_1/sleeteretal.pmd.pdf, accessed 10 June 2010.

Stanley, T. (2002) 'Let's get digital: Using multimedia and the internet in community cultural development', in Adams, D. and Goldbard, A. (eds) *Community, Culture and Globalization*, New York: Rockefeller Foundation.

Thurman, H. 'Howard Thurman quotes', http://thinkexist.com/quotes/Howard_Thurman, accessed 7 May 2010.

Ybarro-Fausto, T. (2002) 'Introduction', in Adams, D. and Goldbard, A. (eds) *Community, Culture and Globalization*, New York: Rockefeller Foundation.

20

BECOMING WITH/IN SOCIAL ECOLOGY

Writing as practice in creative learning

Catherine E. Camden-Pratt

Where to begin?

Coming to this blank page is exciting and terrifying. Exciting, because of the opportunities and possibilities that open up, and terrifying, because within these same possibilities chaos awaits.

In this chapter I take up writing as a mode of becoming. My narrative is woven from glimpses of my sixteen-year engagement with/in social ecology,[1] where, as an educator and researcher I work with becoming with/in social ecology using creative learning. My practice of writing lets the body into the writing, acknowledging affect and the idiosyncratic life paths towards the always in process present, as well as ecological understandings of self as being in relation to others and to the world.

(*How neatly I write this introductory paragraph and yet this neatness belies the complexity and messiness that rests within its borders!*)

Taking up the permissions and opportunities of this blank page, feeling an excitement mixed with terror, I am writing to *know*;[2] to become. I use a variety of text types and narrative voices to signify different contexts, relationships, and learnings and to highlight the diverse ways in which I make meaning. The narrative voices include: a theorising voice with its traces of other scholars; *an emergent reflective voice*, ***an her-storical poetic voice*** and a referenced end-note voice which does not interrupt the narrative flow of the central text. This is ecological writing that acknowledges the relationships and the contexts in which the writing takes place, and their influences on the writer/writing.[3] I invite you into this multi-voiced text.

When I teach,[4] I suggest that students begin where they are. And so I begin this story of creative learning, of becoming with/in social ecology, here at this blank page, sitting at my desk with my excitement and terror:

I sit at home typing, my laptop keys responsive under my fingertips as a lover to my touch. Breath in and out. Heavy rain pounding my tin roof, late autumn sunshine banished. Outside my window a yellowing carpet of leaves, trees half shorn, red against grey. I am cosy, toasty warm. Fresh lemon ginger tea with a dollop of honey beside me. I have sat here looking out this same window watching the seasons change over many years now, years that have seen my PhD researched, written and submitted, my book published, academic chapters written, articles submitted, phone interviews given, lectures and tutorials designed, rich student assessments read, emails received and emails sent. And poems grow. The familiar clutter of coloured postcards, bills to be paid, art gallery stubs, pens, rulers and scissors in large honey jars, favourite rocks on my window sill and images of Dhurga Kali and Frida Kahlo surround and hold me. Notepads, spiralled journals and documents in process wait, clamouring for attention. Bookshelves line the wall above me to the right, to my left my open kitchen with its not-yet-done breakfast dishes by the sink. Behind me, my large red sofa living room of comings and goings, of family gatherings, food and friends, research meetings, art making, a now silent piano, photos of family and friends, paintings and books, books and books. This is a site of my becoming, with echoes of other sites. Here my ecological relationships are revealed – the relationships in and through which my creative learning takes shape. How come after so many years I still feel this fear as I come to the blank page of academic writing? What is it I am scared of? What does it ask of me? What does it promise that means I run the gauntlet of my fears to meet myself/s[5] *and you here?*

Contextualising becoming with/in social ecology

Becoming is always in process. It is a dynamic process in which an individual and a community are able to move away from previous limitations and become something new, which in turn continues becoming.[6] This transformation occurs through relationship. Thinking relationally entails facing and understanding yourself in critical relationship with others, including the human and non-human world – your ecological self. Engaging with social ecology can provide a way to understand, map and deepen conscious becoming by foregrounding the relational self/s. I am not suggesting that social ecology is the only way through which and with/in which to do this, it is, however, a way that I know and apply.

I began studying social ecology in 1994 as a postgraduate student at the Hawkesbury campus of the University of Western Sydney. Now, sixteen years later in 2010, I lecture in social ecology across a number of campuses at the same university. As a social ecologist, I approach life in a holistic way, endeavouring to remember and act with the awareness that my actions affect not only myself, my loved ones and those communities with which I am directly involved, but also those in other local and global communities, as well as future generations.[7] An ecological approach means thinking and acting with environmental awareness, awake to the vast complex web of which humans are a part, and for which humans have immense responsibility.[8] Being a social ecologist means I know that I don't know, that I understand not knowing to be as crucial as knowing. It means meeting my miss-takes as part of me – miss takes which enable my learning. It means theoretically and practically *learning my way towards*[9] those things I do not understand. Being a social ecologist means working within a co-created community, knowing I am part of the whole and the whole is part of me. Applying a social ecology perspective, immediately and critically positions me – and each of us – in mutually affecting social, political, cultural, spiritual and environmental relationships.

Social ecology recognises that individuals with/in community can be agents of change in their own lives as well as in community – and that individual change is not located outside of community. As I experience it and have been privileged to witness over the years in social ecology, becoming – movement away from previous limitations into something new – is enabled by understanding and working with socio-ecological awareness. My practice of creativity takes place with/in these understandings.

A practice of creativity: Writing as creative learning

Creativity begins with an offer and its acceptance.[10] Without this there is nowhere to go. Creativity flourishes when an accepted offer is extended. When this happens there is somewhere new for the creator (e.g. individual, community, class, family) to go. This is becoming, and every stage of this takes place with/in contextually located ecological relationships.

Other aspects necessary to the creative process are preparation, incubation, illumination and verification.[11] (*Reading these words neatly written one after the other in a row conveys an orderliness that is an outright untruth!)* The creative process of writing to know is messy and chaotic, and is one place where the tensions between academic writing and creative learning are experienced. Most academic writing effaces this process so that it is not visible. (*How can I write an academic text here that takes up creative leaning and makes it transparent, so that it can enable becoming for me and others as writers of academic texts?)*

To be in this messy creative flow requires trust, if not trust in *myself* as the authoritative author, then trust in the process. *(The authoritative author is crucial in academia where competitive models thrive; where students and academics are graded; where territories, disciplines, funding models and identities are built through publications in the right journals and books. My tummy tightens and my heart thumps: I am an unstable authority, always in becoming – where can this identity be validated in competitive academic landscapes?)* It helps to have maps of the creative process so that we can recognise ourselves in the landscapes in which we find ourselves.[12] I know from previous experiences that saying 'yes' to and extending on an offer, sweeps me up into a creative process, into unknowing, into the chaos that is integral to creativity, into an exquisite excitement, with fear as ally and enemy.[13] Swept up into the chaos, into the unknowing makes me vulnerable – to myself, the process, and to the knowing that might emerge. It also makes me vulnerable to its verification; by me, by wider communities of family, friends, colleagues, students and unknown readers. Creative emergence is facilitated by knowing how to be with/in chaos fuelled vulnerability. Paradoxically, learning how to be with this vulnerability is learned through being with it.

How did I build confidence with this vulnerability, this chaos? Stilled wind, bright blue sky and winter's chilly sunshine cast shadows outside my window. I search my files seeking reflections on my creative learning through writing, and find a narrative I wrote in 1994 in my first year as a postgraduate social ecology student. It was part of my writing portfolio for the subject Writing Ecologically.[14] *I titled it 'Reflections of a creative amputee', and in it explored what I experienced as the amputation of my creativity through my growing up as a child and teenager in an abusive family and chaotic home. I talked about my feeling the ghost limbs of my creativity using what I'd heard from amputees talking of feeling limbs that were no longer there:*

> ***At almost fourteen the path of a creative life seemed doomed to one or more in a range of options. Take your pick: alcoholism; madness; child abuse; poverty; bottomless never-ending depression; disappointment; ultimately death at your own hands.***
>
> ***Does this seem a digression to you? It is not. It is a placing in context, a telling of how the ghost limbs came into being. For that's how it seems to a creative amputee. I know they are there – these limbs of the poet, who once could run like the wind and soar with wings wide and free – yet i cannot move them. It is the stump of my toes that wriggle. Others look, see movement, and in the illusion of the self that we see in others, believe i am indeed full limbed. Now i am unwrapping those ghost limbs, no longer four years old, no longer fourteen.***
>
> ***Re-membering these limbs of mine bring pain and tears long before there is joy in movement:***
>
> > ***What if i'm no good,***
> > ***discovered to be a fraud at 'it'?***
> > ***Uncovered***
> > ***discovered***
> > ***laid bare***
> > ***all pus oozing***
> > ***and weeping,***
> > ***not able to be cleaned***
> > ***and put back together***
> > ***neatly***
> > ***so that the scar line is negligible,***
> > ***hardly seen:***
> > ***velvet flesh***
> > ***olive,***
> > ***smooth to touch;***

no scar,
no trace of the gaping,
serrated edges.
Is that how it is meant to be – poetry?
Bringing joy to a heart;
scar-line unseen?
Poetry like that i cannot write.
A soul like that i do not have.

Becoming with/in social ecology: Writing as practice in creative learning

Studying with/in social ecology gave me permission to bring my own life into view as a worthwhile and integral focus of study, re-shaping what was possible in my own learning and its re-presentation. Although the politics of representation in research now recognise the autoethnographic voice,[15] as a postgraduate student this permission in an educational setting was recent to me. Feminism meant that the personal was political.[16] Social ecology showed that it wasn't possible to separate the personal from the social, political, spiritual and environmental dimensions of life. Post-structuralism described how discursive practices shape each of us, and how we continue to shape discourses through our own actions.[17] I came to understand how my becoming and the kinds of possibilities I saw for myself were shaped by the kinds of social discourses I had engaged in: education, gender, social class and normalcy, among others. Perhaps more importantly, in an academic context, I began to grapple with the theories that could name, describe and legitimate what I knew from my own experiences. I began to develop agency as I started to consciously unpack the discourses that had shaped my becoming, and to make choices about how to inter-rupt these to enable becomings for myself. *('Choice'. 'Agency'. Tricky words that imply and promise so much, and yet are problematic. How does context and opportunity shape agency and choice? What complexities are hidden in these words, what possible judgements lurk at their edges?)*

Writing with/in social ecology was an opening; an initiation into, and possibility for, agency in terms of imagining otherwise than the problematic and shifting discursive frames within which I lived. I was then able to act differently. Applying ecological understandings I began to better understand and consciously work with how *my* agency had the capacity to also influence others through relationships and community engagement. As a teacher – in schools and later at university – I began to articulate the ways in which I co-created with students the kinds of educational practices which enabled, rather than stifled, agency.

As I re-read 'Reflections of a creative amputee', I remember writing it, the pain I experienced unwrapping my ghost limbs, slowly bringing them into materiality again. With/in social ecology, I learned new ways to write academically, ways that incorporated my own experiences as opportunities for critical reflection on the kinds of discourses that shape/d me. It was this creative learning that I accessed through new forms of academic writing that enabled my becoming.

Within the social ecology framework, creative learning meant using whatever mode of language best opened up and represented the topics I was studying. I painted, wrote narratives and poetry, made installations, wrote plays, wrote and sang songs, accompanying myself on the guitar – all alongside an academic narrative that explored the theory that working creatively had opened up. I came to understand how academic representation is limited by the languages and modes of writing accepted in academia. I saw how working with social ecological understandings meant allowing different kinds of representations/voices, none privileged over another.

I wrestle with the concepts vying for their places on the page. The linear nature of writing belies the complex multilayered holistic embodied process of creative learning and becoming. What do I say, what do I leave out? What makes sense here? Where is the 'So what?' in what I have to say? My own becoming with/in social ecology using creative learning, led me to a social ecology PhD that contained narrative writing, poetry, a play, paintings and collages all layered with/in academic theorising.[18] *Afterwards I felt like I had birthed a multi-headed, multi-limbed Ganesh! Today I wonder at the permissions I made for myself and was supported in by my supervision panel, and am glad for the PhD and subsequent popular press publication and exhibitions that emerged.*[19] *This is the creative learning in and through which I became other than the creative amputee. It is this knowing that supports my teaching. So, how has my becoming within social ecology influenced my teaching? What permissions for writing and other ways of languaging learning, have I opened up for and with the students I teach?*

So what?[20]

I have written in detail elsewhere about my ecological approaches to pedagogy.[21] Here I touch on how writing and languaging 'otherwise' as part of the creative learning in my university classes facilitates student and teacher becoming. My context is the undergraduate subject 'Learning and Creativity', which sits inside an Education Studies Major for students who intend becoming teachers. Before going further it is important to say I make it clear at the beginning and throughout the subject, that students have the right to not participate in tutorial and lecture activities. We discuss this approach – 'right to pass' and its associated responsibilities – as part of an engaged pedagogy.

With the intention of students developing confidence and skills in writing and languaging 'otherwise', I scaffold a variety of arts-based approaches throughout the semester. All lectures use a variety of media, from the verbal and visual to the dramatic and musical as well as a variety of styles; including narrative, theoretical, poetic. In each class and lecture, students are given opportunities for, and support to, write and re-present their learning in new academic forms. In tutorials students write using a 'first thoughts'[22] approach, reflecting on their responses to tutorial-specific content. I also give time in classes and lectures for students to reflect in writing on their learning to date in the subject, and on the questions and curiosities they have about their learning. We use collage, painting and drama as ways of critical engagement with theory, then applying well-scaffolded exercises, we build up a sensory-rich narrative and poetic vocabulary to use in writing short narratives and poetry that critically explore and represent our learning. We – including me as teacher–learner – have the choice to share, or not share, our work with each other in class, building and nurturing a trusting community through the risks we take in doing this. Through these practices students develop increased agency and a greater capacity to make informed choices and take action in ways that support their lives and the lives of others.

Social ecology critically positions students as central to their learning. In assessments, students are asked to critically reflect on their becoming throughout the semester. In written assessments, students are able to and encouraged to write and represent their learning in new academic forms. They are given permission to use 'I' in their written work, while at the same time integrating critical analytic discussion of referenced texts in responding to the essay topic. In the first written assessment in the subject a small number of students take up the opportunity to include poetry and narrative and visual images. However, in the final document *all* students take up this opportunity to write 'otherwise', and in doing so develop their own diverse voices and contribute to growing new academic writing and representational practices. The opportunity to write 'otherwise' facilitates critical thinking and deep personal understanding. I remember 'Rose' and her final document – her personal reflections indicative of the quality of student reflections. Rose wrote about her first assessment where she had used her Aboriginal storytelling style in writing about Stranger Danger. In her final assessment she wrote how she'd realized that in the beginning she saw all of us as strangers and felt in danger and about how she no longer did – it had been okay to be herself and how she was, she'd said 'I guess we've done so much yarning together we're like a family – a new mob.'

Student assessment is often a place of competition rather than a shared celebration of learning. How can assessment become otherwise; become a harvesting of personal and community learning? In the final assessment students creatively represent their own story of learning using media of their choice. We set up a marketplace of creativity with stalls giving simultaneous performances of painting, writing, collage, drama, cooking, music, dance, sculptures and installations. After an opening ritual, students spend time engaging with each other's stalls, talking with each other and the tutors about their work and give written feedback on their peers' presentations. This is social ecology in action in the assessment space, inter-rupting the usual university assessment discourses. It is individual learning as community learning, becoming as individual and collective. It is political. It brings the personal stories into the public space, with students learning from their peers' stories about depression, dyslexia, autism, privileges of a happy childhood, refugee experiences, cultural knowledge, domestic violence, child abuse and neglect. This kind of educational assessment practice, in which stories that are often silenced in university settings are instead legitimated and shared, enables the possibility of social change. As becoming teachers, students experience new ways of doing assessment and validating individual lived experiences, they learn about learning from each other, expand their understandings of each other, of education and of what builds (a learning) community. In these ways they participate in and witness each other's becoming and how creative learning facilitates this.

Ecological approaches to teaching – in which learning is co-created by all participants – place me as teacher alongside learners as a co-learner. I know what becoming can be enabled through creative learning and so my teaching comes from the inside-out and not from the outside-in; that is, I come from experience into theory. Walking the talk is crucial in an educational context, particularly in teacher education where as teachers we are simultaneously teaching content and pedagogical process. I walk my talk. I demonstrate my own creative learning and becoming as an ongoing process and am able to build a substantial meta-narrative on the creative pedagogies I use in teaching. Since my first classes as a university educator I have shared my creative work with students – paintings, research play-scripts and poetry, performances and narratives. In 2010, for example, I published my first poem – Ashes – outside an academic context.[23] Although it was a poem that grew from my research, I entered it into a poetry prize that might lead to publication in an anthology of Australian poetry[24] and I felt a surge of joy, different from my academic publications, when it was accepted. I used this process and product as a current example for students of creative learning through writing, and the becoming that is possible. Through this, students are able to witness the practice I am teaching.

Where to end?

My own becoming is central in and crucial to my capacity and agency as a teacher–learner. I bring myself as who I am into my teaching – as all teachers do. Within socio-ecological understandings I am also critically positioned as a learner, as a person in becoming with/in my ecological relationships. It is to this I return.

Late autumn has become winter and now spring as I complete this chapter. Terror has vanished, and in its place butterfly winged excitement in my belly. I accepted the offer of writing this chapter and entered creative learning and here now on the page is my articulated becoming – the extension of the offer I accepted. It is this realised promise of becoming that meant I ran the gauntlet of my fears to meet myself/s and you here on the page. I take down The Green Fuse from my book shelves and re-read my poem 'Ashes':

I collect you,
grasp you firm in my two hands
and place you
passenger-side,
on the floor
of my white VW beetle.
I hold you
between my sandaled feet.
My children buckled into their car seats
look out the window,
sunshine slipping by.
My husband
at the steering wheel,
eyes on the road.
You are
fixed in a tin
I do not open,
with a label
I do not read.
I know it is you:
your foot to foot rocking,
your frozen eyes ablaze;
your Largactyl perfumed madness
stilled by death.

It is the first poem in Part 1 of a narrative I am writing. The poetic narrative builds on my PhD research, research which built on my postgraduate study as a social ecology coursework student, studies that allowed me to find my voice in writing 'otherwise' in academic contexts, unpacking the discourses that shaped me. I reflect on how the scar-lines of my soul have entered academic and non-academic communities in a number of ways and feel deep gratitude for the offers of social ecology, and the support and guidance I received in accepting and extending on them. My fear of not being good enough in academic writing abates as I write with/in my integrity – scar-lines seen. There is no ending; my full stop is contextual and artificial, determined by chapter size and publication dates. My becoming with/in social ecology: writing as practice in creative learning continues.

Notes

1 I refer to the model of social ecology as first taught at Hawkesbury campus and later at Penrith and Bankstown campus of the University of Western Sydney and referred to in Introduction, Chapters 1, 10 and 11.

2 Richardson (1994: 39) 'writing … is also a way of "knowing" – a method of discovery and analysis … by writing in different ways we discover new aspects of our topic and our relationship to it'.

3 Many chapters in this book demonstrate writing ecologically. Among them are MacKinnon (Chapter 8), Cameron (Chapter 10), Frauenfelder (Chapter 19), Hartlage (Chapter 21), Birrell (Chapter 22), Whelan (Chapter 23) and Gannon (Chapter 27).
4 I teach undergraduate postgraduate and research students in social ecology at the University of Western Sydney.
5 I have used 'self/s' to signify the postmodern notion of a multiple located self that is contextually and relationally located.
6 Deleuze and Guattari (1987) talk about 'becoming' as a verb that is continually enacted, rather than arrived at. I explore and apply Deleuzian becoming in educational settings (Camden-Pratt 2009), with Bronwyn Davies, Susanne Gannon, Constance Ellwood, Katerina Zabrodska and Peter Bansel.
7 Hill (1999); Camden-Pratt (2008).
8 Rozak et al. (1995); Capra (1996); Jasmin Ball and Kathryn McCabe (Chapter 9).
9 I first heard John Cameron (Chapter 10) use this phrase in staff meetings at UWS. I find it a concept full of permission, forgiveness and grounded hope.
10 Pierse (1993). David Wright (Chapter 18) talks about this process in drama. I discussed its application in pedagogy (Camden-Pratt 2008) and in clowning as transformational process (Ellwood and Camden-Pratt 2009).
11 Poincare (1952) initially described these four stages in analysing his process with/in mathematics; Wallis (1976) gave these names to each of the stages that Poincare described. Neville (2005) has an excellent discussion of the creative process and creativity in school classrooms.
12 There are a variety of possible maps for creativity – see Yardley (Chapter 7) for a rich map.
13 I have taken up many offers and extended on them in a variety of ways – writing (poetry, narratives, plays, theorised), painting, sculpture, collages, dramatic performances (Camden-Pratt 2002, 2003, 2006, 2008), Horsfall et al. (2004, 2007).
14 This remarkable unit was taught by Marilyn McCutcheon. As a student I was fortunate to learn within a vibrant women's lineage – Judy Pinn (1984–2001), Marilyn McCutcheon (1984–1996), Debbie Horsfall (1994–2000), Virginia Kaufmann Hall (1994–1996), Karen Bridgeman (1994) – who embodied deeply connected practices of research and pedagogy. Other women who shaped social ecology at Hawkesbury included Frances Parker, Chris Winneke, Helen Kiernan, Lesley Kuhn, Ariel Salleh. I remain appreciative for all I learned from these early women social ecologists. Judy Pinn and Marilyn McCutcheon were central to establishing and developing social ecology at Hawkesbury. As Women Out To Lunch, I collaborate with Debbie Horsfall, Judy Pinn, Virginia Kaufmann Hall and Karen Bridgeman in performing research and publishing new academic writing.
15 Church (1995); Bochner and Ellis (2006).
16 Hooks (1984, 1998) and Lather (1989, 1991).
17 Gordon (1980) edited a selection of Foucault's (1972–1977) interviews and writings, in which discourse and its impacts in a variety of settings is unpacked. Jennifer Gore applies this directly to pedagogy (1993).
18 My PhD, *Daughters of Persephone: Legacies of Maternal Madness,* is available from the University of Western Sydney Library website.
19 I first exhibited my PhD paintings in 2003 and have shown them at conferences, used them in plays and in lectures. In 2006 *Out of the Shadows; Daughters Growing up with a 'Mad' Mother* was published.
20 I learned this important question from Debbie Horsfall, once my PhD supervisor and now my friend, colleague and co-researcher.
21 You will find a detailed discussion of social ecology, creative arts and pedagogy in Camden-Pratt (2008). Some fundamental understandings from my pedagogy are that the classroom is an ecosystem with/in wider ecosystems and teaching ecologically

recognises that learning is simultaneously a personal and a community experience. Ecological understandings acknowledge that every action in an ecosystem has an impact on the whole, so that what may appear for example as a one-on-one interaction between a teacher and student impacts on the whole. Socio-ecological understandings recognise that learning is co-created by all of the participants – teacher and students – and all they bring visibly and invisibly to the learning. My practices draws from a lineage that includes Freire (1996, 2005) and bell hooks (1994, 2003).

22 See Natalie Goldberg (1986, 1990).

23 See Gannon (Chapter 27) for a rich discussion of and demonstration of poetic inquiry.

24 Camden-Pratt (2010).

References

Bochner, A. P. and Ellis, C. (2006) 'Communication as autoethnography', in J. Gregory et al. (eds) *Communication as Autoethnography: Perspectives on Theory*, Thousand Oakes, CA: Sage.

Camden-Pratt, C. (2002) *Daughters of Persephone: Legacies of Maternal Madness*, PhD thesis, Australia: University of Western Sydney.

Camden-Pratt, C. (2003) *waiting to be re-membered; Paintings and artist conversation*, exhibition, Blue Mountains Women's Health Centre, Katoomba NSW.

Camden-Pratt, C. (2006) *Out of the Shadows: Daughters Growing up with a 'Mad' Mother*, Sydney NSW: Finch Publications.

Camden-Pratt, C. (2008) 'Social ecology and creative pedagogy: Using creative arts and critical thinking in co-creating and sustaining ecological learning webs in university pedagogies', in Transnational Curriculum Inquiry http://nitinat.library.ubc.ca/ojs/index.php/tci, accessed 9 February 2011.

Camden-Pratt, C. (2009) 'Relationality and the art of becoming', in B. Davies and S. Gannon (eds) *Pedagogical Encounters*, New York: Peter Lang.

Camden-Pratt, C. (2010) 'Ashes', in C. Williams (ed.) *The Green Fuse, The Picaro Press Poetry Prize*, Warners Bay, Australia: Picaro Press.

Capra, F. (1996) *The Web of Life: A New Synthesis of Mind / Matter*, London: Flamingo.

Church, K. (1995) *Forbidden Narratives, Critical Autobiography as Social Science*, Newark, NJ: Gordon and Breach.

Deleuze, G. and Guattari, F. (1987) *A Thousand Plateaus; Capitalism and Schizophrenia*, London: Athlone Press.

Ellwood, C. and Camden-Pratt, C. (2008) 'Becoming Blossom, becoming Oddbod: Clowning as transformational process', in B. Davies and S. Gannon (eds) *Pedagogical Encounters*, New York: Peter Lang.

Friere, P. (1970 / 1996). *Pedagogy of the oppressed.* London: Penguin.

Friere, P. (2005). *Education for Critical Consciousness*. London & New York: Continuum.

Goldberg, Natalie (1986) *Writing Down the Bones: Freeing the Writer Within*, Boston, MA: Shambala.

Goldberg, Natalie (1990) *Wild Mind: Living the Writer's Life*, New York: Bantam.

Gordon, C. (ed) (1980) *Foucault Michel (1972–1977) Power/Knowledge. Selected Interviews and other Writings*, London: Harvester Press.

Gore, J. (1993) *The Struggle for Pedagogies*, London: Routledge.

Hill, S. B. (1999) 'Social ecology as future stories: An Australian perspective', *A Social Ecology Journal*, 1, 197–208 University of Western Sydney.

Hooks, B. (1984) *Feminist Theory, From Margin to Center*, Boston MA: South End Press.
Hooks, B. (1994) *Teaching to Transgress*. New York: Routledge.
Hooks, B. (1998) *Wounds of Passion*, London: The Women's Press.
Hooks, B. (2003) *Teaching Community: A Pedagogy of Hope*, New York: Routledge.
Horsfall, D., Bridges, D., Camden-Pratt, C. and Sammon, L. (2004) 'A performance of difference', *Journal of Reflective Practice*, 5(1): 91–110.
Horsfall, D., Bridgeman, K., Camden-Pratt, C., Kaufman Hall, V. and Pinn, J. (2007) 'Playing creative edges: Performing research – women out to lunch', in J. Higgs et al. (eds) *Being Critical and Creative in Qualitative Research*, Sydney NSW: Hampden Press.
Lather, P. (1989) 'Deconstructing/deconstructive inquiry: The politics of knowing and being known', paper presented at the American Research Association Annual Conference, San Francisco, 27–31 March.
Lather, P. (1991) *Getting Smart: Feminist Research and Pedagogy with/in the Postmodern*, London: Routledge.
Neville, B. (2005) *Educating Psyche: Emotion, Imagination and the Unconscious in Learning*, Melbourne: Flat Chat Press.
Pierse, L. (1993) *Theatresports Downunder*, Sydney: Improcorp Australia.
Poincare, H. (1952) 'Mathematical creation', in B. Ghislen (ed.) *The Creative Process*, Berkley, CA: University of California Press.
Richardson, L. (1994) 'Writing: A method of inquiry', in N. K. Denzin and Y.S. Lincoln (eds) *Handbook of Qualitative Research*, first edition, London: Sage Publications.
Rozak T., Gomes M. and Kanner D. (eds) (1995) *Ecopsychology: Restoring the Earth, Healing the Mind*, San Francisco, CA: Sierra Club Books.
Wallis, G. (1976) *The Art of Thought*, New York: Harcourt and Brace.

PART 4

ECOLOGICAL STORIES

21

WE ARE WHAT WE EAT

Christy Hartlage

> The food crisis is not a political, but a spiritual one, prompted by a breakdown in our relationship with the Earth.
>
> (Amrein, 2010)

Every morning my partner and I make breakfast together. When the weather is cold we usually have hot porridge or polenta or pancakes. Every evening I make dinner for my family. I use as much organic food as possible: eggs from our chickens, fresh, whole, seasonal foods. We sit down together and eat. It is not easy to do this. Small children are sometimes tired and do not want to be distracted from play and stories. Parents are sometimes tired and would rather have takeaways for dinner, and sometimes we do just that. However, I want my children to know what good food tastes like, and to experience how their bodies feel when they eat well. I want my family to be nourished with the delicious flavors and sensations of the Earth. I want there to be time in our days that is set aside for sitting down together and talking to each other. For myself, I want to pay attention to what I put in my mouth so that I can pay better attention to how I use the resources of the Earth.

We are what we eat. What we take into our bodies, literally, becomes who we are. Our decisions about food actually, materially, shape our bones and our bellies; they give us strength and vitality, and so shape our personalities. What we put in our mouths is who we are. 'We literally take the environment into ourselves and merge with it … Because it utilizes the senses, eating, more than any other human experience, brings us to our fullest and most intimate relationship with the environment, (Kimbrell 2010: 24).

Whether we choose to eat a home-cooked meal or McDonald's, whether we eat local, organic vegetables, or we indulge in imported Beluga caviar, our thoughts and choices about food create our cells, the texture of our skin, the shine of our hair, and the sparkle in our eyes. We make meaning of the world around us through the community that is fed around the table. Conversations, prayers, arguments, rituals and celebrations that both teach and create culture, family stories, politics, and our understanding of our families and friends take place with and around food. To keep ourselves alive we spend an extraordinary amount of time and energy thinking about food. However, food preparation, like so many ordinary, daily chores, is not valued. I have never heard anyone say that they don't have time to make money. Yet we often say that we do not have time to cook, sometimes we don't have time to eat, or we rush through meals to get on to more important things. How can we help but become detached from the places and people around us if we cannot pay attention to how we feed our bodies? If we recognize that through ordinary activities such as growing and preparing food we make meaning of the world in a multitude of ways, we can begin to recognize the transformative potential of sitting down to eat.

Making dinner is not separate from thinking about the kind of world I want my children to grow in. Because we must eat, we must also be able to understand ourselves better by understanding our relationships to food. When I think about the kinds of nourishment that my family and I need as a whole people, I think about feeding our intellectual, spiritual, emotional and physical elements. Eating is about sustaining and nurturing our connection with the living community around us. It is said that the way someone cooks for you is the way they love you. The way we share food expresses the fullness of relationship that we enjoy with all living beings around us; recognizing this pleasure we can be satiated with a deeper understanding of the world. 'The Vimlakirti Sutra says: "When one is identified with the food one eats, one is identified with the whole universe; when we are identified with the whole universe we are identified with the food that we eat"' (Curtin 1992: 127).

When we conduct ourselves with this sort of relationship with food how do our lives change? How are we in relationship with the Earth if we are in this kind of deep relationship with what we put in our mouths? 'Despite of the necessary time we spend feeding ourselves, worrying about calories and nutrition – we pay astonishingly little reflective attention to our food' (Heldke 2003: xxvii). We don't think about where our oranges have come from, or what the life of the chicken was like. I think this is the point where we can begin to recognise the breakdown in our relationship with the Earth. We think about the number of calories in the foods we eat, the amount of fat or protein we consume, but we do not think about the quality of the air and soil where the orange grew, or whether the chicken lived in a light grassy paddock or a factory farm. Understanding the qualities of the places where our food is raised may help us to understand ourselves in relationship with the wind and air, the soil, the trees, and the animals who, like us, depend on the Earth for sustenance.

> It seems to me that our three basic needs, for food and security and love, are so mixed and mingled and entwined that we cannot straightly think of one without the others. So it happens that when I write of hunger, I am really writing about love and the hunger for it, and warmth and the love of it and the hunger for it … and then the warmth and richness and fine reality of hunger satisfied … and all is one.
>
> (Fisher, 1983: 353)

There is pleasure in relationship. Understanding that our relationship with the Earth is our primary relationship: the relationship that keeps us alive, can lead us to a sensual relationship with our natural community. Fostering an understanding of ourselves in relationship with and dependent upon the places that feed us can nourish a meaningful identification with cycles of growth and regeneration. The varied and timely sensuality of food brings us into a comm(on)union with nature and community that has met our need for pleasure over centuries. We nourish, sustain and comfort ourselves in response to the pleasures of others and in balance with nature. This is a need that is much more complex than physical survival. 'Eating with the fullest pleasure', Wendell Berry (1990: 153) suggests, 'pleasure, that is, that does not depend on ignorance – is perhaps the profoundest enactment of our connection with the world.' Eating is one of the most ordinary activities we take part in, and it can also be one of the most sensual and enriching. Identifying ourselves with the world requires us to understand, in our bones and in our flesh, who and what we depend on for sustenance. Understanding our dependence on and participation with the natural world through the cycles of food, nurtures the ecological consciousness of who and how we are in a community of living beings.

When we develop a deep, responsive relationship to a place we become sensitive, sensualized to the ordinary movements, rhythms and pleasures of the place. Nourishing consciousness of the cycles of food could be one way for a society to learn its way to place responsiveness (Cameron 2001). To experience the sweet tangy acid of biting into a fresh tomato, skin still warm from the sun, we must be able to observe and respond to the tomato plant in its place. Our interactions with the soil and the plant during its life nourish an intimate appreciation of its flavors and textures so that we respond fully to the moment when juice slides down our throats and seeds spill onto our chins. Because the pleasures of ripe food can be experienced only briefly, they require us to be responsive to a place in this moment. The time and attention we must give to a place in order to notice the swelling of pea pods and the advance of the rosy blush on the skin of a peach enables a knowing of place that leads us to respond, with excitement and anticipation, to the ripeness around us. Knowing the cycles of food production in a place is a way of honoring the place and our connection with it. Wendell Berry suggests that 'eaters … must understand that eating takes place inescapably in the world, that it is an inescapably agricultural act, and that how we eat determines, to a considerable extent, how the world is used' (1990: 149).

Understanding the cycles of nourishment, the care/full and timely cultivation that goes into creating a really good meal is a juicy element of place responsiveness. Meals that take time also make time for nourishing conversation and relationships with family and friends. This is when we really understand that our hunger is not simply a need for calories and protein, but it is hunger for connection, relationship and love. When we better understand how what we put in our mouths connects us to the lives of cows and date palms and earthworms, we will also better understand our longing to be in relationship with the natural world that includes the people around us.

I think that understanding starts with noticing. Simply noticing that the water from a certain stream tastes different from the water that comes from the tap in my kitchen, and different again from the water in the river near my favorite hot pool, gives me information about this place and how my body responds, practically and sensually to it. To feel myself really part of a place – and the place a part of me, I need to know when the asparagus season begins, and where I can find the sweetest strawberries. The adventure of tasting a new place by finding out about the edible things that only grow there, sensitizes me to the elements in the soil, the frequency of rain and the direction the wind blows around me. Tasting a place, feeling how it moves in my mouth and fills my belly, makes the place a part of me, always. We can express a richness of connection with the land by learning to taste and feel the textures of ripeness around us, in our bodies and on our tongues. In a culture where most of the food we eat is picked and refrigerated and shipped before it is ripe it has become a luxury to eat really fresh food. Think about what that says about our ecological relationships. What we put in our mouths, what creates the substance of our bodies, is not ripe, juicy, at the peak of sweetness. Surely by denying ourselves the pleasure of food grown nearby and picked when it is at its best we deny an opportunity for another kind of knowing of place. We are denying ourselves the sensual flavor of the places where we live. However, if our choices about food take into account our ecological relationships, we can express our pleasure and love for the Earth, by sharing food with the living community around us.

References

Amrein, Elizabeth (guest editor) (2010) *Resurgence Magazine*, 259, March/April.

Berry, Wendell (1990) '*The Pleasures of Eating.*' *What are People for?,* San Francisco, CA: North Point Press.

Cameron, John (2001) 'Place, belonging and ecopolitics: Learning our way towards a place responsive society', *Ecopolitics*, 1(2): 18–34.

Curtin, Deane W. (1992) 'Recipes for values', in Deane W. Curtin and Lisa M. Heldke (eds) *Cooking, Eating, Thinking: Transformative Philosophies of Food*, Bloomington, IN: Indiana University Press.

Fisher, M.F.K. (1983) *The Art of Eating*, London: Picador.

Heldke, L. (2003) *Exotic Appetites*, London: Routledge.

Kimbrell, Andrew (2010) New Food Future?, *Resurgence*, 259, 21–4.

22

SLIPPING BENEATH THE KIMBERLEY SKIN

Carol Birrell

Prologue

The skin of the Kimberley is tough. It's old and rough and scaly like the crocodile whose presence is all over land and sea up this way. The stifling heat, the crumbling ancient red rocks, the pure aqua water, the flies chasing the sweat on your back, the smells and the stories, are all held by the skin of this country. To slip beneath this skin is no easy matter. You have to nick it with a sharp knife, then you are in. Or, more likely, the Kimberley has now got in under your skin.

I invite you to enter deeply into this Aboriginal country and into indigenous ways of being. 'Country' is the word used by Aboriginal Australians denoting land, usually their own particular piece of land holding stories and traditions that go back millennia. In meeting country, we are meeting ways of knowing and meaningful coincidences that from this perspective are not coincidences at all, but emerge from a world inundated by spiritual significations and profound intelligences.

Aboriginal Elder Bob Randall (2003) asks if white Australians could go beyond the conceptual to an experience of an 'inner eye' where one could 'feel our situation, to read people, to talk to country'. Aboriginal Yuin Elder and Senior Lawman Uncle Max Dulumunmun Harrison (Birrell 2007) speaks of 'goin' in [to sacred sites] with whitefellas eyes and comin' out with blackfellas mind'. Stanner (1979) called it 'thinking black'. Is it possible for an outsider to become an insider? What does this take?

If reconciliation between Indigenous and non-indigenous people in Australia today is to mean more than the token gesture, more than a hollow 'sorry', surely an in-depth engagement with Aboriginal culture on its own terms is required. If one desires to sit comfortably with this land, surely one needs to surrender to the land on its own terms.

This essay reflects one part of my twelve-year experience of moving towards an understanding of Aboriginal culture as an outsider. It is the narrative of a white woman invited onto country by a Kimberley (north-west Western Australia) Elder. It reveals a sense of surrendering into the potency of the land and into another way of being. It is no easy journey.

The smell of fear

Invitation offered; acceptance given. Entering in to the unknown. No information except we would be out on a boat for about two weeks. A whiff of fear. A whiff of excitement.

Imagine this: five people on one small pearling boat, named 'Badmarra', sea eagle. One true saltwater person – four to shortly become saltwater persons. Treacherous waters that demand explicit knowledge and finely honed intuition. Perishing heat that radiates off ancient crumbling cliffs, offset only by cool, ocean, on-board drafts. Aqua water. Crocodile country. Barramundi Dreaming country. Sandflies, mosquitoes and mangroves that ooze them forth onto white unaccustomed flesh. Night skies where the spaces between the stars become as important as the stars themselves. Deep silences.

This is the Kimberley. Wild inaccessible country registered on many sections of maps as 'unsurveyed' and 'extremely remote', yet known like the palm of your hand by traditional owners. One of these is Donny Woolagoodja, an artist responsible for painting up Wandjina images in caves on his country. He's a Worrora man and it's his invitation that got me here.

Donny is healthy compared to most of his mob, despite his diabetes. He lives in town with his white partner who tends to his health with a ferocious attachment matched only by her house cleaning obsession. When going out on the town he returns to his cattle droving past by dressing up in blue Levi jeans, Cuban-heeled riding boots and black cowboy hat. He sticks his hands in his front pockets, thumbs out, and sways on his heels. Even now, that era long gone, he cuts a mean figure that turns many heads, black and white. 'Yo!', we exclaim when we see him in all his finery. But it is a 'Yo!' tinged with sadness for a past life of pride and guts and vitality. Donny knew who he was then.

One night camped on the beach, I awake in the night to see Donny walking down towards the water, the huge tide moving out, a fullish, very bright moon stillness. I wonder if he can't sleep, is restless, hot, or maybe checking on the boat. Next morning I ask him if the boat was okay during the night. He looks blank. I tell him how I'd seen him walking down to the water in the middle of the night. He denies it all, so too does his partner. 'Might be my spirit', he replies, matter-of-factly.

It is new for me to explore country by boat, to get to know sea as country. As varied in detail and story as any landscape. I begin to see the dimensions of this world, both in the pattern of its ever-changing surface and what is deeply below. My sense of smell becomes more refined as I smell stiff salty clothing, my body odour, the strong pungent reek of mangroves at low tide, and the distinctive briny smell of a close crocodile. I have grown sea legs in a relatively short space of time. From a beginning of trepidation, with clutching white knuckles on handrails and squirmish stomach, I begin to feel exhilarated by the sea. I tan a golden brown, lose worry lines and salt encrusts my whole body like a second skin. I ride comfortably on the boat as she bumps uncomfortably across the waves. Even my chronic constipation, no doubt caused by on-board exposure to many eyes other than crocodiles, dissipates somewhat.

We catch and eat our own fish, either raw as sushi or cooked over the hot coals of the open fire on the beach. I am taught how to gather oyster shells, to prise them off the rocks then to shuck them open to discover the quivering palm-sized delight inside. No oysters ever tasted so good coming from these pristine waters! After about four or five days, the land alters with faces peering out of the depths of ironstone clad cliffs. It's like a switch turned on. I wonder if I, too, have been shucked by this country? No force here, though, more a willing gentle unfolding, the crack widening little by little, my soft pulpy flesh exposed and the country slowly showing its soft pulpy flesh.

There are tracks everywhere on the beach, but I manage to make out dog tracks, large dog paws moving right by our tents just above the high tide mark. I guess maybe the fishing people keep dogs. I tell Donny and he informs me it is dingo, a lone male whose female pack waits under cover until he calls. We see a very large croc crossing the bay right then but I can't pick up his scent like I could the other day when we were in the mangroves and Donny picked him up. I'm anxious. Well, if it's not the croc then it will be the dingo that gets me. Donny says if I'm scared to just stay up all night by the campfire, stoking the logs. 'That'll keep them away'. 'Great', I say, then fall asleep shortly after.

Old bleached bones

I am not sure when mention was first made of the baby's bones. I had already moved into that timeless realm where days were indistinguishable from tidal fluctuations. Early on, I think, and when we were close to the cave that housed the bones. Donny spoke of how tourists would trek up to the cave to check out the bones. He wanted to put an end to this, this lack of respect. There was a need to put these bones to rest he said, and to put this spirit to rest. There is a deep silence from the rest of us. I ask no questions even though I am intrigued: 'Who is this baby? Where did she come from? How did she get to be in this cave? Why did she die?'

At first a story, then an intention, then a plan. It was spoken of each day like a punctuation, in the same way the tides punctuated our days and made themselves pivotal to our lives. The bones would be moved. Threads of relevance would drop into conversations as easily as dropping a fishing line over the edge of our boat. There was an issue of logistics and when to go according to the tides, of course, of how to get access, and to which cave they should be transferred, the bones, that is, and who should go. But even though there was this presence invoked each day, I wondered if the actual shifting of the bones would really take place, or if this narrative would ebb and flow in its own way, with no clear resolution. It did not seem to matter, whatever the outcome.

One breakfast on the beach, Donny asks if we have had any dreams the night before. I immediately recall one of me dreaming through the experiences of others who have lived in this country. I seemed to flow rapidly through their entire lifetimes on this Kimberley land. The experience was as if their lives were concertinaed and passed into me, a type of compression of life files that flooded my being and bestowed me with the depth and breadth of their experience of this landscape. A whole range of people like pearlers and cattle men and townies and Aboriginal people, the one thing connecting them all being a passionate connection with this country. Donny offers no comment, just sits fiddling with a stick in the sand, backwards and forwards, making his own tracks.

There is a protocol to follow if one is to pay respect in visiting 'Namaralee', the Wandjina that Donny has often painted in his art works. This is to wait close by for three days before attempting to get to his cave. Donny says 'Namarali' is the greatest spiritual warrior to have ever walked the earth. I believe it. The last time Donny went to the Namaralee cave and did not follow the protocol, he and the group wandered the bush completely lost for at least eight hours. The cave refused to be found. Donny laughs when he tells this story. I am sure not laughing.

The three days have felt just right. There has been a build-up and always in my mind has been this intention: patience and respectful waiting. It is like Namarali senses our presence and we his. We are accompanied by a few people who are staying in a fishing camp close by. Two are children, one very young, who will have to be carried most of the way. Somehow, these people have materialised to join us. And again I surrender to whatever arises, disallowing feelings of intrusion. There is early talk of paperbark collection, for the baby's bones.

The eyes of the Wandjina

We are smoked on the beach before departure. Donny grabs some branches, piles them one on top of the other and waits for the green leaves to send out their smoke. He tells the story of that old Wandjina and how he ended up in that cave. This is a story that echoes through the cliffs as it does through generations of blackfellas. A story held by country in the forced absence of those who should be hearing it.

Black eerie rock 'Wandjinas' check our credentials and allow us through.

Badmarra sits breathing deeply on an aqua ocean; red, black and gold flag flapping … waiting.

It is not an easy walk. Heat that makes you sweat from the minute you begin to walk, thousands of flies that have been waiting to leap upon sweaty bodies from the moment you step one foot away from the sea, hard unforgiving red rocks that pulsate more heat into an already intolerable atmosphere, and spiky spinifex shrubs that announce themselves with a sharp stab. All this, and a pervading stillness. Donny strides ahead, a lizard-like ease over rocks. I don't see him brushing off stinging green ants as the frequent shouts indicate the rest of us are doing. It's not just the kids who find it tough going. By the time I have almost had enough, we are there, in the welcome coolness of a well lit cave.

I breathe very slowly as Namaralee espies me, and I him. His eyes follow me around the cave. There is no escaping them. I take to the floor, looking up at him, eyes ablaze. And I immediately know why I am here. The others chat, eat, drink and take photos. I so want utter silence but this is not possible, not here, not now. I notice a dragonfly present. When I finally, reluctantly, come out of this prostrate position, the children are painted in red ochre. I want this, too.

There is a qualitative shift once Donny has applied the ochre to my face. It is some deepening, of being more fully present, of sinking into country even more. He invites me to rub the blood red rock myself and put some on my arms and hands, while laughing about the fact that this ochre will take about three days to wash off. I take the small piece of well worn ochre in the palm of my hand. It is red and glistening and fits there as if it was meant to. Namarali watches me closely.

I rub the ochre on the larger rounded slab of rock bearing the stain of thousands of ochre rubbings, soft at first then harder, in order to get more of the ochre off into a thick paste. I then smear this onto my body, arms first, then legs. It looks like dried blood on my legs. I look down at my hands, now stained in red, the lines of my palm etched deeply with colour.

Not breathing

It is a real wrench to leave the cave, a physical pain in my body. We split into two groups: the fish camp people are to go a different way and collect the paperbark. I have a foreboding and am glad I am with the Donny group.

I am elated now, after the visitation, and we make it back to the beach quicker than the way in. Hot and tired and glad to plunge in the water just briefly, knowing there is a big croc in this bay. I don't rub the ochre off, wanting it to soak right into me, through my skin, into my muscles, my bones, my marrow.

The other group is not back. We stay still and wait. There is a little concern. And we wait.

Badmarra is still flapping, a lot closer now, or so it seems, in high tide.

Hours go by. And still they have not returned. We think they are lost. Donny is unconcerned. The air is becoming more and more oppressive. Flies multiply, body drips with red stickiness and flimsy trees offer little shade. We wait and wait. This is like the world has stopped, stopped on the in-breath.

Finally they arrive, hot and flustered, very obvious they have been through an ordeal. One child is really distressed, probably heat exhaustion. One man, the leader of the group, has a huge pile of paperbark under an arm. All are overheated and stare at the water. They stare even more strongly at us, paddling in it to keep cool. They are rattled, that's for sure, but no one tells us what happened, other than the fact they got lost.

We take the paperbark back with us on Badmarra. It sits right at the front of the boat with bits flying off as we pick up speed. I keep staring at it, imagining the wrapping of the bones. The distressed girl lies on the floor of the boat, fussed over.

So we are one step closer to this next enactment, yet I still have no idea if it will stop here or continue on to its supposed climax.

Piercing the flesh of country

The baby bones have to be retrieved first, at high tide. Who is to go and who is not to go is the topic of conversation. It seems at first discussion that no women and children are allowed. Then all that changes, and the men will collect the bones in the morning, on the high tide, and then we will all take them to the new cave. I know this may change too and let it all wash over me, with no attachment to outcomes again.

The day of the bones. Or is it? Plans made abruptly change and we are washing at the fishing camp. Women washing while the men go off to retrieve the bones. The young boy cries as he really wants to accompany them. We sit and wait their return, still unsure what may unfold that day. When they do return around lunch-time, I see the paperbark jutting out of the front of the small dinghy. I can't take my eyes off it, like trying to avert your gaze when you pass a road accident, and know it is serious, and you still stare all the more. There is talk of the smoking that will be needed: us and the little dinghy and Badmarra. I gather this is to clear that dead child's spirit from us all. It makes sense to also cleanse the boats, but it still takes me by surprise.

The readiness to leave seems to jump up rapidly, as if the tide has changed and we all have to go … now! And we are off, following Donny across the jagged rocks, sweat pouring off in the first few minutes of shady tree departure, flies gathering by the score. The fishing camp man closest to the front has the bundle of paperbark bones over his shoulder. It is tied up and around with red string. As we walk in single file, funereal like, there is no speaking. We are respectfully taking the bones home, laying the spirits to rest. In such a hushed atmosphere, we could easily hear Badmarra's flag flapping, but the wind, too, has dropped and there is no flap; the cloth falls limp and still. All sorts of stories begin to take form in my mind, connecting these bones. The strongest one, which will not go away, is that the baby was murdered by whites.

I don't get bitten by green ants on this day. Quite remarkable, really. They are still prolific but for some reason I am not attacked.

The bones are laid on a ledge in a cave in a new burial place. The paperbark is removed. I go up close and see that the baby bones are wrapped in a cloth, so small and vulnerable. I am full of grief. It sits in my throat and won't budge. Somehow, I know it is more than the death of this baby.

We are smoked good and proper on returning to camp, the small boat too. Badmarra will have to wait until the next morning, the food removed, the decks scrubbed down. Big fire, big smoke, big story. The sun setting is spectacular. We are all hushed again, sucking in the smoke deeply. This is a good thing that has been done. I keep on moving to follow the smoke, determined to cleanse myself inside out.

Old skin, new skin

The tides are big and strong, heading towards the next full moon, when I am back in Sydney. Well, not so big and not so strong as I have known.

My skin is peeling since I got home. Not just flaky little pieces, but big chunks of thick skin that I help remove with sharp nail scissors. I have never seen my skin come away like this.

I place my Namaralee painting on the wall of my apartment with his eyes facing out to the big northern skies.

Badmarra flaps in the wind, waiting.

There was another invitation issued. An acceptance. An honouring.

'I am "Gadja", grandmother to Donny. Skin group "Jinkurn". "Wororra" woman. Saltwater woman.'

I call this out not once, but many times, hoping it will be heard not only here, in the place I live, but heard way up there in the Kimberley, way up there in that cave, sniffed out by that old Wandjina spirit with the eyes that go right through you, to your soul, and hold you, suspended in another world, another time.

Note

This chapter is drawn from my Social Ecology PhD thesis (Birrell 2007) that examined the intersection between a Western and Indigenous (Australian Aboriginal) sense of place.

References

Birrell, C. L. (2007) *Meeting Country: Deep Engagement with Place and Indigenous Culture*, PhD thesis, University of Western Sydney.

Randall, B. (2003) *Songman: The Story of an Aboriginal Elder of Uluru*, Sydney: ABC Books.

Stanner, W.E.H. (1979) *White Man Got No Dreaming: Essays 1938–1973*, Canberra: Australian University Press.

23

CLIMATE ACTIVISM AND TRANSFORMATION

James Whelan

Sufi fable describes a phoenix-like bird, the *huma*,[1] that lives its entire life high above the earth, where it feeds, interacts, reproduces and lays its egg; an egg that hurtles toward the earth while, inside, the unborn chick begins to form. Fran Peavey – activist educator, storyteller and author of *Heart Politics* – compares the huma chick's dilemma with that faced by social justice activists. 'Will the chick's beak harden in time to peck its way through its shell?' Fran asked during her workshops. 'Will its wings form and strengthen in time to avoid collision and to soar back to the stratosphere?'

The fable and Fran's allusion describe the crisis lived by climate activists.

Climate change is the problem of our generation: a challenge that tests us individually and collectively and demands, 'Do we have what it takes to avert catastrophic change?' The challenge has triggered the emergence of a global climate movement, the most visible and urgent contemporary social movement. Community members who form this movement face a steep learning curve and time is racing by.

I am inspired by the determination and resourcefulness of grassroots climate activists and daunted by their learning challenge. Many come to the movement, and to activism, with minimal political experience. They find themselves part of a rapidly changing social and political landscape where policy debate is polarised, decisive interventions are obstructed by powerful vested interests, and where the consequences are catastrophic and irreversible.

Like the embryonic chick hurtling towards earth, the climate movement is transforming and building power to meet the demands of the times. To the extent that a movement has consciousness and can reflect on itself, it shares the chick's crisis: what are the chances of averting catastrophe? Is it possible?

As a climate activist in the 1990s, I couldn't imagine how the social movement would evolve. Although climate science was already well established, few environmentalists were engaged in the push and pull of climate policy. When we did, we lacked the power that the movement now draws from widespread awareness and concern. More than 90 per cent of Australians now accept that human activities are contributing to elevated levels of ambient CO_2 that is warming the earth. In the 1990s, we weren't ready to mobilise, and instead researched, prepared submissions, presented arguments to committees and attempted to persuade politicians. And we had minimal impact. Greenhouse gas abatement policies were adopted but nothing prevented an exponential growth in emissions. Under both major political parties, Australia's climate policy favoured polluters and facilitated spiralling emissions.

But times are changing. Today's climate movement is dynamic, diverse and powerful. It spans the political spectrum from small 'L' Liberals and technophiles to deep ecologists and socialists. Climate activists seek to influence policy development and implementation by state and Commonwealth governments, but increasingly reject reformist tactics in favour of community mobilisation and civil disobedience.[2] They have learnt from observing the older environmental non-government organisations (ENGOs) whose cautious diplomacy has reached the limits of its effectiveness. As Sun Tsu, the sixth-century BC Chinese military strategist, said, 'Tactics without strategy is the slowest route to victory' (Sun Tsu 2005). Throughout the Australian climate movement, activists now believe that confrontation and crisis are essential moments in the creation and enforcement of laws to curb emissions.

In mid-October 2009, I received a curious email. The subject: 'Permaculture workshop invitation.' I'm interested in permaculture, but wasn't looking for workshops to attend. My finger hesitated over the 'delete' button while I briefly skim-read the email. There was more to it. A climate change 'activity' was being organised in Canberra. Did I want to come? The email didn't say much more than that, but promised an information package would be sent by post. I replied that I'd be keen to come. When the information package arrived, I learnt that the workshop was a pseudonym for a climate change protest at Parliament House. The timing was excellent. The Rudd Government's Carbon Pollution Reduction Scheme (CPRS), which had been widely rejected by the climate movement, was due to be discussed by the Senate and an Australian delegation was preparing to head to Copenhagen for the global climate negotiations. The protest could draw attention to shortcomings of the CPRS and communicate support for a strong Australian position in the Copenhagen Conference of Parties.

In Brisbane where I lived, a gathering was quickly organised for activists who planned to head to the national capital. Along with the familiar faces – people I knew from local action groups and events such as the annual Walk Against Warming – were several new folks. Although many had only heard of the 'permaculture workshop' a few days before the Canberra protest, they were ready to travel 1,000 kilometres and potentially be arrested in an action they knew little about. Not because we were a hardened direct action crew – many people had never participated in civil disobedience before – but because we *needed* to act. This was equally true of climate protests in Brisbane during 2009. Newcomers to the movement were quick to commit to courageous and potentially arrestable actions. A recent recruit to climate activism, Fiona, participated in her first demonstration in October when she paddled a kayak to the Brisbane River coal-loading facility while two young activists locked on to the conveyor belt. A month later, Fiona was prepared to be arrested.

I joined one of several cars leaving Brisbane. During the long drive south, we learnt about each other's motivations. What drew us to climate activism? What did we hope to achieve? If the action achieved its objectives, what would change? Each of us had our own answers: we hoped to inspire and give courage to others; to 'send a message' to politicians who weren't listening and create political space for policy development; to shake things up. What of civil disobedience? Why was this the chosen tactic? During the previous two years, strategists in the movement had hypothesised that a breakthrough in climate policy would require mass civil disobedience on an historical scale. Working backwards, that would require each action to build from the preceding one and for arrestable non-violent protest to be considered by activists and the wider community as both necessary and acceptable. As long as arrestees could be cast as marginal and extremist, our claims would be easily dismissed. The permaculture workshop organisers shared this analysis and asked us to wear formal clothes for the protest to present a dignified image that would resonate with television viewers.

The climate movement took a step toward normalising civil disobedience in July 2008 when activists from around the country gathered in Newcastle for our first climate camp. The week-long convergence culminated in two days of non-violent direct action to prevent coal trains from reaching the world's biggest coal port. I helped develop a survey asking participants in those protests what had motivated their participation in direct action. Before climate camp, 73 per cent said they were likely or very likely to engage in direct action. After a long and tiring day that saw 150 arrests and several instances of police violence including punched faces and broken bones, this rose to 93 per cent. Almost all the protesters were more inclined than before to participate in arrestable protest actions. Two hundred people were ready to put themselves on the line and were firmly convinced that civil disobedience was justified. Organisers had created a family-friendly, 'citizen' face for the protest and a supportive and inclusive decision-making camp culture, helping activists learn the power and potential for direct action.

After fifteen hours of driving, we reached the outskirts of Canberra. The organisers of the permaculture workshop had arranged our overnight camping at an actual permaculture property. In the golden afternoon sunshine, we set up our tents, greeted old friends and made new ones. With a growing sense of anticipation about the next day's action, people drifted off to their sleeping bags.

Before dawn, we woke to pack our tents and board buses for a briefing session at the Australian National University. Pouring into the lecture theatre, it was clear that the workshop invitation had struck a chord. More than 200 people arrived from around the country: 'apolitical' neighbourhood action group members, renewable energy enthusiasts, professional campaigners and radical grassroots activists with widely ranging politics, class backgrounds and ages, united in their sense of urgency. The doors were locked and our sense of anticipation grew. One of the organisers asked us to cluster with people who knew us and could vouch for our involvement in climate activism: a clever device to identify any infiltrators.

Recognising faces in the crowd, I felt a strong sense of connectedness. This was my tribe and we were ready for action.

The briefing explained the objectives, tactics, contingency plans, roles and media strategy for the day's action. Many of us had participated in a peaceful protest at Parliament House in February when more than a thousand people held hands to encircle parliament, so we had an inkling what might be planned. This time, we'd be blocking the public access entrance to demand that the Rudd Government commit to a 40 per cent reduction in greenhouse emissions by 2020 in Copenhagen. Since his acclaimed endorsement of the Kyoto Agreement, the prime minister had back-pedaled on climate change and was now proposing a modest 5–15 per cent emission reduction target. We discussed some responses we might encounter, the image of our non-cooperation with the police and our physical presence. With banners, placards and spokespeople communicating our demands, we agreed that a dignified silence would convey our determination and unity or purpose: we wouldn't chant or shout. To optimise the media and political impact, it was important that many of us were willing to be arrested. The organisers asked us to for three lines: a 'red' queue for those willing to refuse to move on and risk arrest, 'orange' for people who would cooperate with police instructions and 'green' for participants who would disperse once requested. Roughly equal numbers joined each queue initially, then there was a drift from green to orange and orange to red as people reassessed their readiness for arrest.

Just after 8:00 am, we re-boarded the buses and rode to the car park beneath Parliament House, filed silently off the bus and up the stairs to the entrance atrium. Without speaking, we sat close together in rows, gradually filling from the glass doors until our 200 bodies filled the antechamber. The entrance was impassable. A red-faced security guard tried to haul one of the group away but was quickly persuaded this was futile. Parliament House was shut down. Tourists queuing to visit the parliamentary gallery were bemused. Several asked protestors to explain our action.

I looked from person to person. Who did I know? When had I last seen them? What were they experiencing? Were they OK? Where did the new faces come from? What were their stories? What brought them here? Every so often, our eyes met and held. It seemed we were all looking for something. Connection? Recognition? Maybe checking for fear or discomfort? There was a strong sense of care. I looked for Holly, a young community organiser who'd progressed through student politics to rapidly become a significant networker and mentor in several progressive social movements. Her organiser instinct is strong, so I wasn't surprised to note that she was scanning the group. Her crew, many of whom she had recruited and mentored. Without words, she asked how Bill was. A war veteran in his eighties, Bill was rapidly becoming a climate protest veteran. He stood silently at the back of our throng, holding a handwritten cloth sign which read, 'We expect leadership at Copenhagen Mr Rudd. This is time for real action on climate change.' Did he need a chair? A hand? Knowing Holly, I imagined she'd connected with Bill at other times and may have helped bring him into the movement, just as she'd build bridges with unionists and traditional owners.

The media arrived and began interviewing our spokespeople. Then 30–40 police. All men. At 11:37 am, a police inspector gave us a fifteen-minute warning to move on or risk arrest. Senators started emerging from the building to interact with us. Seeking media exposure, dialogue or conflict? Greens Senator Christine Milne first, saying all the right things. Yes, that's why we're here. Then climate sceptic Senator Bill Heffernan with his incoherent hostility. Our discipline held. Don't be baited. Ssssshhhh. In silence, we shared food, glanced around to connect non-verbally.

Senator Stephen Fielding emerged to challenge us. A climate sceptic with a laughable prop – a placard with a graph apparently depicting global temperatures over the last twenty years. 'It's not warming! The science is inconclusive.' And still they came. Senators Barnaby Joyce and Bob Brown. Both ends of the spectrum – a dinosaur and a visionary.

The police shuffled in along one wall of the antechamber, pushing back the protestors just far enough to form a solid line of their own.

One of the protestors rose. Jenny has spoken for us many times. A recent and (she said) reluctant activist, Jenny spoke from the heart and was quick to tears. 'This is unfair. Many of us are here for our kids.' After attending a meeting in her Sydney suburb, Jenny quickly found herself convening a new climate action group and helping build the national grassroots movement. The connection between Jenny's commitment to her children and to climate action resonated for me. I'd taken my sixteen-year-old son to climate camp. Louis had witnessed the police dogs, charging horses, arrests and his father's commitment. It brought us closer.

The ground was hard. I grew increasingly conscious of my bones as the hours ticked by. My legs yearned to stretch. I worried how it must be for Bill, sitting just along from me, and listened to his laboured breathing as he struggled to rearrange his legs between our tightly packed bodies. As the police took out each row of uncooperative protestors, their line moved ever closer. Our sweat from the long journey, camping and collective anxiety was confronted by the chemical uniformity of their boot polish, dry cleaning and deodorant. The strange silence stretched out, swirling silence, sense of uncertainty and unexpected calm. Would it hurt as they lifted and dragged? Would the anger and fear I saw in their eyes give way to aggression? I swallowed. My dry mouth. Craving water, food.

The arrests began just after midday. After the long wait, we watched closely. Would there be violence? Pain-inflicting holds on our wrists and necks? Their instructions were scripted: 'You have been instructed to move on. If you do not, you will be arrested ... You are under arrest. If you do not stand and walk we will lift you and you may fall forward.' Several people complied before the police met one of our 'red' group. As Susie was lifted and carried, her head bumped a sandstone doorway hard. Then she was out of sight, surrounded by blue uniforms. Many of us knew Susie for her patient determination and commitment to non-violence. A seasoned community activist, Susie had mentored many environmental activists.

Quickly, a routine was established. In response to their script, we replied, 'I will not cooperate with a government unwilling to act to prevent catastrophic climate change.' One by one, activists were led, carried or dragged away, their faces communicating fear, conviction, concern and solidarity.

By 1:30 pm, the police line was against me. I tried to establish eye contact but saw only a row of grim faces. I couldn't interpret their feelings or thoughts. Many wore dark or mirrored and impenetrable glasses. In some faces, I thought I saw doubt. Guilt?

Then it was my turn. The scripted interaction, being carried to the stairs, identified, charged, loaded into the paddy wagon and taken down hill to join the growing crowd outside old Parliament House. The arrests continued for more than three hours.

The egg accelerates downward, reaching terminal velocity.

As it plummets, the chick anticipates and dreads the collision ahead. Each day, the case for climate action strengthens. A 1°C increase in temperature will result in 10 per cent loss of all known species, yet policy dialogue is framed in terms of a rise of 2–3°. Already, Pacific Islanders and coastal communities are being displaced. Feedback mechanisms are accelerating change. Predictions of biophysical change are being overtaken, yet the climate sceptics and fossil fuel lobbyists are standing in the way: Christopher Monckton, Ian Plimer and Bjørn Lomborg facing off against 1,300 of the world's leading scientists as if on equal terms.

To crack the eggshell and emerge ready for flight, we need to be outward looking: to shift from self-doubt to determination.

The climate movement invests much of our energy in analysis, self-analysis and critique. What is our strategy? What tactics are most likely to achieve our objectives? Are they the right objectives? How will change happen? The movement's responses to these questions indicate division more than diversity. Some non-government organisations believe climate catastrophe can be averted through dialogue with government and industry and through modest increments in climate policy. A 5 per cent emission reduction *is* a step in the right direction. The Australian Conservation Foundation, Climate Institute, Australian Council of Social Services and Australian Trades Council of Unions, united as the Southern Cross Alliance, endorsed the proposed CPRS. They were condemned by the grassroots in equal parts for supporting an inadequate policy and squandering the political space that had been created through protests and arrests. Some activists constrain their actions and language to the space they perceive to be acceptable, referring to polluters as 'trade exposed emission intensive industries'. Greens' candidate Ben Spies-Butcher describes this as 'debating like a government in exile'. Other parts of the movement believe climate justice cannot be assured without fundamental shifts in governance and power. Demanding a 'clean energy future' without dismantling corporate power, according to Rising Tide North America (2010), is like 'putting lipstick on a corpse'. Similarly, the 50,000 participants in KlimaForum, a civil society gathering that ran parallel to the Copenhagen COP (Conference of the Parties), endorsed a declaration calling for 'system change – not climate change'.

Participatory governance and global democracy, more than reductions in industrial emissions, is the focus of the World People's Conference on Climate Change and the Rights of Mother Earth convened by Bolivian President Evo Morales in Cochabamba, Bolivia this April: 'for the people to decide'. Increasingly, activists recognise that the movement must have broader and more political objectives than short-term climate policy: to wrest power from corporations, redistribute opportunity, transform governance and rethink democracy.

This isn't the first time we have plummeted earthbound. We've flown before. In 1987, 196 member states of the United Nations endorsed the Montreal Protocol on Substances that Deplete the Ozone Layer. Heralded by Kofi Annan as 'perhaps the single most successful international agreement to date', the protocol came just fourteen years after chemists began studying the link between chlorofluorocarbon (CFC) molecules and ozone depletion.[3] Attributing the Protocol's adoption to scientific research and political negotiation overlooks the significant role played by civil society groups. The ozone layer is now protected despite sceptics, scientific uncertainty and powerful economic interests. The climate movement can draw insight and encouragement from CFC and other toxics campaigns.

There are solid grounds for optimism. Climate change is routinely on top of the policy agenda. Policies are being developed and implemented, commitments are being made. Despite widespread disappointment about the outcomes of Copenhagen, the gathering was historical. More than 100 national leaders participated; the prime ministers of Canada and Aotearoa (New Zealand) reluctantly agreeing to attend in response to civil society campaigns. More than 22,000 registered observers sought to hold delegates accountable during the negotiations and the world witnessed the movement's diversity: insiders and outsiders, pragmatists and zealots. Civil society groups educated and exposed, petitioned and pilloried their national leaders. The backlash led to their ejection from the forum, a reminder of Frederick Douglass' truism, 'Power concedes nothing without a demand. It never did and it never will' (Douglass 1985).

And the demands are coming thick and fast. New modes of networking allow the movement to mobilise locally and globally through viral actions such as the Global Work Party on Climate Action on 10 October 2010 that involved more than 7,000 community events in 188 countries. As Harald Zindler, a founder of Greenpeace observed, 'The optimism of the action is better than the pessimism of the thought' (Greenpeace).

We need to believe in our capacity for flight. Having pecked a hole in our shell, we can't now doubt our capacity and must do what we can while we can. Fear of success and its shadow, belief in movement failure, Bill Moyer (1990) observed, is almost endemic in social movements. It's natural to feel overwhelmed by the magnitude of the required changes and the odds stacked against us. But we won't spread our wings as long as our imagination dwells on the alternative and we argue about tactics.

Notes

1 http://www.absoluteastronomy.com/topics/Huma_(mythology), accessed 9 February 2011; http://www.co-intelligence.org/y2k_inbetween.html, accessed 9 February 2011.
2 See for instance Coal Swarm, http://www.sourcewatch.org/index.php?title=Portal:Coal_Issues, accessed 9 February 2011.
3 Richard Benedick suggested the climate movement could draw lessons from the Montreal Protocol campaign in 1991 – almost twenty years ago.

References

Benedick, R.E. (1991) 'The diplomacy of climate change: Lessons from the Montreal Ozone Protocol', *Energy Policy*, 19(2): 94–7.

Douglass, F. (1985 [1857]) 'The significance of emancipation in the West Indies', speech, Canandaigua, New York, 3 August 1857; collected in a pamphlet by author, in Blassingame. J.W. (ed.) *The Frederick Douglass Papers. Series One: Speeches, Debates, and Interviews*, *Volume 3: 1855–63*, New Haven, CT: Yale University Press, 204.

Greeenpeace http://www.greenpeace.org/australia/admin/image-library2/the-optimism-of-the-action-is, accessed 15 February 2011.

Moyer, B. (1990) *The Practical Strategist: Movement Action Plan (MAP) Strategic Theories for Evaluating, Planning and Conducting Social Movements*, San Francisco, CA: Social Movement Empowerment Project.

Rising Tide North America (2010) 'The Climate Movement is Dead: Long Live the Climate Movement', http://risingtidenorthamerica.org/download/rtna_climatemovementisdead.pdf, accessed 20 March 2010.

Sun Tsu (2005) *The Art of War*, Boston, MA: Shambhala Publications.

24

WE ARE NOT ALONE, THE SHAMANS TELL US

John Broomfield

We're in a spot of bother. Humanity, that is. I know it, you know it and our students know it. For us as teachers to disregard the problems facing the Earth and to proceed with business as usual at any level of education would be a betrayal of trust. Nor will it suffice simply to produce elaborate proofs that there is a global problem. Our students want to know how to make a difference. They need hope. And it won't come if all we can offer is another scientific theory or technological fix. In large part we're in this mess because of an over-reliance on scientific rationality and its technologies.

> Reason sets the boundaries far too narrowly for us, and would have us accept only the known – and that too with limitations – and live in a known framework, just as if we were sure how far life actually extends … The more the critical reason dominates, the more impoverished life becomes … Overvalued reason has this in common with political absolutism: under its dominion the individual is pauperized.
>
> (Jung, 1961: 302)

Of course science will offer some valuable new directions, but at the same time we must expand our vision to seek non-scientific alternatives. To make a difference, we must search for different understandings.

I am fortunate to live in a country, New Zealand, where many of my compatriots have an understanding of past and future that is fundamentally different from the prevailing 'Western' view. Most in our civilization consider it self-evident that we stand facing the future with the past behind us, but traditionally for New Zealand Maori it is the future that is behind them. They stand facing the past and their ancestors, who are a living presence in spirit. It is the vision of the ancestors that guides the present generation into the unseen future, with one clear and overriding purpose: to prosper the generations yet to be born.

> Ng*a wa o mua* 'The days of the past to which we are coming.'
>
> (Maori proverb)

Let us take our cue from Maori and consider the vision of our own ancestors. No matter what our ethnic background, we will discover that our ancestors (except some of the most recent) believed, like Maori, in the existence of spirits. They also stood in awe of the rich diversity of life forms, and they believed there is mutual interdependency between these forms, humans included, given that everything that exists is alive and conscious. They were of the opinion that intelligence is not restricted to humans but is possessed by all creatures – plants as well as animals – and, for that matter, by the Earth itself. Rock, soil, stream, ocean, wind, air, sky, the stars – all are imbued with consciousness.

Recognizing that the Earth and many of its creatures vastly pre-date humanity and are therefore possessed of much older wisdom, our forebears honoured selected landforms, trees, plants and animals as their ancestors. They understood that there is deep wisdom in the rhythms of the Earth and an infinite variety of life experience stored by our fellow creatures and by spirits. Human health and welfare were understood to depend on tapping into this wellspring of wisdom. On a planet that is everywhere alive, conscious and inspirited, humans were believed to have many wise allies for counsel and aid.

What is the relevance of this to our current concern about the fate of the Earth? If the 'star billing' given by us moderns to our species is unwarranted – if *sapiens* (wisdom) is not exclusive to *homo* (humanity) – then could it be that the fate of the Earth is not exclusively or even primarily in our hands? By our ancestors' measure, we have grossly exaggerated our self-importance in the intricate web of life. Is it not conceivable that among our intelligent companions on this whirling voyage through space are some who may be capable of restoring the balance we humans have disturbed, of undoing the damage we have wrought? Possibly there are many more shoulders sharing this burden than we think.

Some of the strongest of those shoulders may be the smallest, as was demonstrated dramatically in the aftermath of the 2010 Gulf of Mexico oil well explosion. As millions of barrels of oil poured unchecked into the ocean from the uncapped well, there was a scramble to devise human technologies that would mitigate an environmental disaster of colossal scope. It took months before the flow was stopped, but in the meantime it was discovered that petroleum-eating bacteria had flourished in the oil plume and contained a vast amount of it. The micro-organisms had not only multiplied at an astounding rate, they also had ramped up their own internal metabolism to digest the oil efficiently. They formed a natural clean-up crew capable of reducing the amount of oil in the undersea plume by half every three days (Hazen et al. 2010).

We may take hope from the fact that this kind of help is available, but we must also start paying attention, as did our ancestors, to what our travelling companions have to say to us. Every ancient society developed communication with the natural world and with spirits, and they had specialists skilled in the techniques of that communication. These women and men were held in high regard, but they were approached with trepidation, because they were perceived to be communing with mysterious and awesome forces. In Old French they were called 'sorcier', those in touch with the 'Source'. The Anglo-Saxons spoke of the 'Ways of Wyrd' known to 'wyzards' and 'wytches'.

Shamanism is the term now applied to what has come to be recognized as a worldwide phenomenon whose practice can be found as far back as we can go in human history. Given the association in the popular imagination of the term shamanism with 'native, tribal' cultures, it will come as a surprise to many to learn that their own ancestors practiced shamanism. We are all descendents of shamanic peoples.

Research over the past 150 years by scholars of comparative religion, prehistory and anthropology has revealed strikingly close similarities in the shamanic techniques employed in ancient cultures and in modern indigenous societies worldwide (Eliade 1972; Cowan 1993; Campbell 1993; Metzner 1994; Vitebsky 2001; Narby and Huxley 2001). The word shaman is borrowed from one of those contemporary indigenous societies, the Tungus of Siberia. We are fortunate there are native shamans still at work, despite the sustained, and in many cases brutal, efforts of colonial governments, Christian churches and medical authorities to suppress them. In the past forty years there has also been a Western revival of shamanic practice inspired by indigenous teachers and reinforced by the recognition that these ancient spiritual traditions are our shared inheritance (Harner 1980; Achterberg 1987; Walsh 1990; Ingerman 1991; Narby and Huxley 2001; Vitebsky 2001).

What do shamans do? They work to maintain or restore harmonious balance between humans and the rest of nature through powerful connections with spirit helpers. This requires a mastery of the techniques of *journeying*.

A shamanic journey is a trance state purposefully induced by a mind-altering activity such as rhythmic movement or repetitive sound, most often steady and sustained drumming. Less commonly, a psychotropic substance is ingested. In their altered state of consciousness, using disciplined techniques individuals can experience visions of flying or entering into the Earth. On their journeys, participants ask animal or guardian spirits to appear and help in finding the answer to a question about their life or about someone else who has requested aid.

Healing is the primary shamanic work. This includes healing of the Earth and its plants and animals. It also includes human healing, both the healing of dissension in groups and of physical and emotional illness in individuals. In the shamanic worldview, *dis-ease* is understood to result from loss of connection to the spirits of nature and consequent loss of soul – individual or collective.[1] Shamanic journeys take us to places where we can recover fragments of lost soul.

Journeying is useful for a wide range of practical purposes, and the experience can be powerful, often surprising the beginner with the cogency and helpfulness of what is revealed. Here is a personal example. Buying property is tricky at the best of times, but when you have been living in America for thirty years and would like to find a place in your home country, New Zealand, it's a major challenge. That's how it was for me in 1991, and I needed help. I received it from a guardian spirit, an eagle. In a shamanic journey, the eagle took me flying over the Marlborough Sounds and showed me a remote property in such detail that I was able to draw a sketch map: the position of the house in relation to two garden plots; the boat shed; the jetty; the shape of the bay. My wife Jo and I brought the map with us when we came to New Zealand three years later. We found a place listed at the first real estate office we visited, and when we were taken to the land, we knew within ten minutes it was the place to which my eagle had flown me. We had no need to look at other properties.

I have another story of shamanic success in real estate. I once participated with thirty others in a shamanic journey to look for a new campus for the California Institute of Integral Studies, the small San Francisco post-graduate school of which I was then president. Many participants found themselves led by their spirit guardians to one particular city neighbourhood. Three people in the journeying group described ornamentation on the outside of a building. One went down a chimney and saw a room with a polished wooden floor and an oriental rug. Another person reported a delicious aroma of baking. Most amusingly, some in the journeying group remarked on a pervasive smell of marijuana in the area. Little wonder. Three weeks later, we found an excellent property half a block from the corner of Haight and Ashbury streets! As we were to discover, the nearest shop, just two hundred metres from our new campus, was a deli, whose baked goods would become favourites of students and faculty, and the journey details of the ornamentation on the building, the chimney, and the room with the polished wooden floor and oriental rug all proved equally accurate.

As this suggests, shamanism can be fun! Shamans are theatrical. In order to rivet the attention of participants, shamans typically wear dramatic costumes and display colourful talismans as they burn herbs and rhythmically whirl, stamp, clap and drum loudly. Almost all of the physical senses of the participants are engaged. As teachers, we should acknowledge shamans as exemplars of excellent educational practice. People learn most forcefully from forms that engage more than their intellects. They remember best what they do, rather than what they read or are told. Effective education must have a large experiential component, and shamanic practice can be a totally engaging experience.

Shamans may be playful, but they are not playing games. Their work has a serious purpose: the evocation of powerful spiritual forces. Shamanic practitioners, as a consequence, must assume responsibility for the welfare of the individuals and groups they guide. As with psychotherapy and similar practices that may bring to awareness deep subconscious memories arousing strong emotions, shamanism must be practiced with disciplined restraint and ethical integrity. Also, with humility. 'In shamanism (as well as with other forms of healing) it is not the shaman who does the work', shamanic counselor Sandra Ingerman observes. 'Shamans are just the instruments through which the power of the universe works. Therefore, asking the spirits for help and trusting that they will be there is the basis of the shaman's responsibilities. Remember, an instrument cannot play itself' (Ingerman 1991: 63).

Effective education must have a large experiential component, I said. Given the current critical imbalance between humans and other species, nature should be a primary area of experiential education. We should balance the abstractions of our classrooms with experiences of the wholeness of living, growing, wild things. Following the centuries-old practice of shamans, students and their teachers should spend time in wilderness to restore direct awareness of the intricate interconnections that sustain life. Quiet time spent away from the elaborate constructions of our cities can help us gain the stillness in which we may hear nature's voices.

Shamanic journeying also can lead to an intimate acquaintance with nature. In his book *The Adventure of Self-Discovery*, psychotherapist Stan Grof reports that in the journeys he and wife Christina direct,[2] many participants experience 'complete and realistic identification' with animals and plants and are given extraordinary knowledge of organic processes. In this mode of consciousness, 'it is possible to gain experiential insight into what it feels like when a cat is curious, an eagle frightened, a cobra hungry, a turtle sexually aroused, or when a shark is breathing through the gills'. This can lead to profound new understandings. 'Subjects have reported that they witnessed botanical processes on the subcellular or molecular level' and had 'experiences of plant consciousness' (Grof 1988: 52–3, 58–9).

Grof commented that to speak of plant consciousness might seem 'fantastic and absurd … to a traditional scientist' (Grof 1988: 59). He was writing in the late 1980s when biology was dominated by molecular geneticists, who, at the time, were supremely confident that all biological function was programmed by DNA sequencing. In the subsequent twenty years, however, there has been a conceptual revolution in genetics and cell biology, with the recognition that cellular networks in organisms are dynamic systems responding intelligently to changing external conditions, even modifying the structure of DNA where necessary. In his 2005 book, *The Biology of Belief*, cell biologist Bruce Lipton writes:

> each cell is an intelligent being that can survive on its own … These smart cells are imbued with intent and purpose; they actively seek environments that support their survival while simultaneously avoiding toxic or hostile ones. Like humans, single cells analyze thousands of stimuli from the microenvironment they inhabit. Through the analysis of this data, cells select appropriate behavioral responses to ensure their survival. Single cells are also capable of learning through these environmental experiences and are able to create cellular memories, which they pass on to their offspring.
>
> (2005: 37–8)

On the basis of such path-breaking research, Fritjof Capra concludes: 'The organizing activity of living systems … is mental activity … Mind … is immanent in matter at all levels of life' (Capra 2002: 30; see also Fox Keller 2000).

We have already observed that this perception of universal consciousness is the crux of the shamanic worldview. By entering the eagle's keen eye, the bear's great strength, the herb's healing power or the flame's searing heat, the shaman shows us passageways to the spirit wisdom of natural forms. Shamans are shape-shifters, teaching that the boundaries between forms are not as impermeable as they may seem. Dramatically, this ancient knowledge that 'there is no wall between species', rejected for three centuries by reductionist Cartesian science, has been rediscovered in this decade by molecular biologists. Lipton again:

> Recent advances in genome science have revealed [that] living organisms … actually integrate their cellular communities by sharing their genes. It had been thought that genes are passed on only to progeny of an individual organism through reproduction. Now scientists realize that genes are shared not only among the individual members of a species, but also among members of different species. The sharing of genetic information via *gene transfer* speeds up evolution since organisms can acquire 'learned' experiences from other organisms. Given this sharing of genes, organisms can no longer be seen as disconnected entities; there is no wall between species.
>
> (2005: 44–5)

'It seems that every process in the universe that one can observe objectively in the ordinary state of consciousness also has a subjective experiential counterpart' in altered states (Grof 1988: 62). This observation by Stan Grof suggests an important reason for the inclusion of shamanic practice in the educational curriculum. Shamanism gives working access to an alternative technique of acquiring knowledge. Although a pragmatic, time-tested system, it makes no claim to be science. Its strengths and limitations are different from those of the sciences and thus complement them. Being affective and subjective, shamanism offers another way of knowing.

In this it serves as shock therapy for students who have grown up with the unexamined belief that modern science is the only true path to knowledge. They have been taught that the scientific method is of a different order from all other human systems of understanding. The claim is that science, and only science, provides a clear window on reality and has the ultimate capacity to answer every question about nature. These assertions are untenable. Modern Western civilization's representation of reality is limited like that of every other civilization. The sciences are cultural constructions to help us get by in the world. 'A scientific theory is just a mathematical model we make to describe our observations', cautions Stephen Hawking. 'It exists only in our minds' (Hawking 1988: 139). Science is a simplification of the universe, which in its unfathomable vastness is always threatening to overwhelm the limited capacity of the human organism to comprehend. 'I suspect there could be life and intelligence out there in forms we can't conceive', observes Martin Rees, British Astronomer Royal. 'It could be there are aspects of reality that are beyond the capacity of our brains' (Rees 2010).

Nonetheless, science reigns supreme and blinds most of our students, like the vast majority of us, to the diverse and richly varied paths to knowledge offered by other civilizations, contemporary and historic. 'Today, the doors of the faerie hills remain sealed against us, for we keep the eyes of our mythic consciousness shut equally tight, refusing to allow cracks to appear in the walls of our present, desacralized world-view.' The writer is Mara Freeman, whose field is Celtic and British folklore. 'Few of us dare to open what W.B. Yeats called the "flaming door" and explore the power that crackles on the thresholds of our reality structures. But to do so might send a revitalizing current through the wasteland of our culture.' Traditionally, Freeman says, it was shamans who had the courage and skill to throw open the 'flaming door'. 'Those skilled in walking between the worlds knew how to harness the power of the threshold where the normal rules of time and space hang suspended' (Freeman 2000: 45–51).

Shamans are edge-walkers and shape-shifters, who dispel the illusion that all is fixed and orderly and controllable.

> A stone's throw out on either hand
> From the well-ordered road we tread,
> And all the world is wild and strange;
> Churl and ghoul and Djinn and sprite
> Shall bear us company to-night,
> For we have reached the Oldest Land
> Wherein the powers of Darkness range.
> (Rudyard Kipling)

Shamanism is an acknowledgment of the awesome spiritual powers that shape the universe. It is an acknowledgement that mystery will remain despite all our science and scholarship. Let us encourage our students to delight in the permanence of the unknowable and to sit in reverence and awe before the majesty of the mysterious. Let us encourage them also to hear the message of the shamans that the moving force in the universe is spirit, which makes life possible and gives it meaning. The exhilarating news the shamans bring is that we are not alone. On a planet that is everywhere alive, conscious and inspirited, humans have many wise allies for counsel and aid. We should lay to rest our exaggerated fears that we do not have the resources to keep this show going. Equally, we must learn humility. The hubris of *Homo sapiens* in claiming superiority over all other species has been the source of severe damage. Humanity is merely one spirit form among countless billions.

> The smallest indivisible reality is, to my mind, intelligent and is waiting there to be used by human spirits if we reach out and call them in. We rush too much with nervous hands and worried minds. We are impatient for results. What we need … is reinforcement of the soul by the invisible power waiting to be used … I know there are reservoirs of spiritual strength from which we human beings thoughtlessly cut ourselves off.
>
> (Ford, 1926)

Notes

1 I would note that although we retain the word 'dispirited' in modern English, it now has little of the gravity it had for our ancestors.
2 Building on their earlier pioneering research with psychedelics, the Grof's have developed a technique they call *holotropic breathing* to induce powerful altered states.

References

Achterberg, J. (1987) *Imagery in Healing: Shamanism and Modern Medicine*, Boston and London: Shambhala.

Campbell, J. (1983) *The Way of the Animal Powers: Historical Atlas of World Mythology*, Volume 1, San Francisco, CA: Harper & Row.

Capra, C. (2002) *The Hidden Connections*, London: Harper Collins.

Cowan, C. (1993) *Fire in the Head: Shamanism and the Celtic Spirit*, San Francisco, CA: Harper.

Eliade, E. (1972) *Shamanism: Archaic Techniques of Ecstasy* (trans. by Trask, W.R.), Princeton, NJ: Princeton University Press.

Ford, H. (1926) *Detroit News*, 7 February.

Fox Keller, E (2000) *The Century of the Gene*, Cambridge, MA: Harvard University Press.

Freeman, M. (2000) 'The flaming door,' *Parabola*, 25(1).

Grof, S. (1988) *The Adventure of Self-Discovery: Dimensions of Consciousness and New Perspectives in Psychotherapy and Inner Exploration*, Albany, NY: SUNY Press.

Harner, M. (1980) *The Way of the Shaman*, San Francisco, CA: Harper.

Hawking, S.W. (1988) *A Brief History of Time from the Big Bang to Black Holes*, New York: Bantam.

Hazen, T. et al. (2010) 'Deep-sea oil plume enriches indigenous oil-degrading bacteria', *Science* 330(6001): 204–8.

Ingerman, S. (1991) *Soul Retrieval: Mending the Fragmented Self*, San Francisco, CA: Harper.

Jung, C.G. (1961) *Memories, Dreams, Reflections*, New York: Pantheon.

Lipton, B.H. (2005) *The Biology of Belief: Unleashing the Power of Consciousness, Matter and Miracles*, Carlsbad, CA: Hay.

Metzner, R. (1994) *The Well of Remembrance: Rediscovering the Earth Wisdom Myths of Northern Europe*, Boston and London: Shambhala.

Narby, J. and Huxley, H. (eds) (2001) *Shamans Through Time: 500 Years on the Path to Knowledge*, New York: Tarcher/Putnam.

Rees, M. (2010) 'Aliens may be staring us in the face', *Daily Telegraph*, 22 February, http://www.telegraph.co.uk/science/space/7289507/Royal-astronomer-Aliens-may-be-staring-us-in-the-face.html, accessed 22 February 2010.

Rudyard Kipling's Verse: Inclusive Edition, 1885–1918 (undated) London: Hodder & Stoughton.

Vitebsky, P. (2001) *Shamanism*, Norman, OK: University of Oklahoma Press.

Walsh, R.N. (1990) *The Spirit of Shamanism*, Los Angeles, CA: Tarcher.

25

THE RELIGION OF ECONOMICS

John Seed, with David Wright

Introduction

Four decades ago, Hazel Henderson wrote that she became an economist to find out 'where the bodies were buried' (personal communication 28 June 2010). I believe the cemetery she is seeking has been well and truly dug over. Though the stink of decay is all pervasive, the residue remains largely hidden from the general public. Here I will attempt to marshal the evidence revealed by a multitude of grave robbing thinkers and sound the alarm. I believe that citing, summarising and propagating these findings is of utmost importance.

I write from the perspective of an environmental activist not a trained academic and certainly not a trained economist. I write out of a commitment to change based on a profound respect for ecological sustainability, as against than the maintenance of the status quo at any cost. I write out of passion rather than equivocal detachment. This is my greatest strength as an activist. It is also my greatest vulnerability. I am vulnerable because I care and because my care extends beyond self-interest.

For the last thirty years I have been working on the conservation of nature. With colleagues, I founded the Rainforest Information Centre in northern NSW, Australia, and have since worked in numerous projects protecting rainforests in South America, Asia and the Pacific. I have lectured on eco-philosophy and conducted experiential deep ecology workshops around the world. With Joanna Macy and Pat Fleming I co-authored the 1988 deep ecology workbook *Thinking Like a Mountain: Towards a Council of All Beings*. The title of this book suggests the deep ecology challenge.

During this time I have been troubled by the irrationality, perhaps even insanity, that is destroying the biological fabric from which our lives are woven. Although actions, led in part by the Rainforest Information Centre, to protect the Australian rainforests between 1979 and 1986 led to a stream of national parks in the Australian states of NSW, Tasmania and Queensland, for every forest protected in those years, thousands were lost worldwide. It has become clear to me that there is no way to save the planet one forest at a time. We need now to address the underlying psychological or spiritual disease that allows humankind to imagine that we can profit from the destruction of our own life support systems. This is akin to, in the words of James Lovelock, 'the brain deciding that the liver is the most important organ in the body and starting mining it' (personal communication 8 July 2010). Or, as Paul Ehrlich says, 'sawing off the branch that we are sitting on' (personal communication 7 August 2010).

I see this problem as more than an environmental or a political problem. I see it as a psychological problem, and believe that the best understanding of the psycho-spiritual dimension of the environmental crisis we are facing is to be found in Deep Ecology. This work, which extends a social–ecological analysis into an empathic identification with natural systems was first articulated by Arne Naess (1973). Naess challenged the illusion of the assumed separation between humans and the rest of the natural world. This illusion is reflected in the underlying power of anthropocentrism: the idea that human beings are the crown of creation and the instrument for the measure of all being. From this perspective anything else can only have instrumental value: it can only be seen as a resource for human beings. Anthropocentrism is a consequence of the illusion of separation, and this separation is entrenched by the ongoing power of the illusion.

Arguably, the roots of anthropocentrism lie in the Judeo-Christian tradition, where 'man' is seen as created in God's image. This suggests that only humans have a soul, and as the creation of god on Earth are entitled, indeed enjoined, to subdue and dominate nature. If we dig at the foundations of classical economics we discover it too has Judeo-Christian roots: nothing has value until humans add their labour and intelligence to it. This means that the Earth is dirt till we dig it up and turn it into value. Bringing the two together: only when the miracle that is 'man' transforms Earth into commodities does the Earth acquire worth. This requires that any proposal to protect nature is almost inevitably 'uneconomic'?

This economic cost–benefit analysis decrees that the benefits of laying Nature to waste trump the costs because, in an extraordinary feat of transubstantiation, such things as air, water, soil and life itself are judged worthless while social fictions such as money are considered real and essential to human experience. This is a 'miracle' of breathtaking power compared to which the parting of waters, or turning water into wine pale into insignificance.

The methodology for the negotiation of this relationship is 'the market'. It has become a knowledge system. As Harvey Cox wrote, in an article in the *Atlantic Monthly* (1999), any

> new grand narrative must enable people to understand the relationship between the stories to which they define themselves as individuals and the stories by which groups constitute themselves and define their goals, ranging from families, local communities, organizations and discursive formations, to nations, international organizations and humanity as a whole.

However, only a deeply religious faith allows us to ignore the absurdity of perpetual growth on a finite planet. The market is the tool of that faith. As Norman O. Brown wrote, 'we no longer give our surplus to God; the process of producing an ever-expanding surplus is in itself our God' (1959: 261).

False theology

Most students of the religious phenomenon of economics see neo-classical economics as a false theology and I will provide some excerpts from some of their analyses in the section that follows. But first I will introduce the curious case of economist Robert Nelson of the University of Maryland, who in the aptly named *Economics as Religion* (2001) celebrates the religious aspect of his discipline.

Challenging the view of economics as a social scientist and economists as social scientists, Nelson argues the role they play more resembles that of theologians. 'Economic efficiency has been the greatest source of social legitimacy in the United States for the past century', he writes, 'and economists have been the priesthood defending this core social value of our era' (2001: xv).

Nelson's metaphor is more than a neat linguistic turn. It reflects a deep-seated social assumption, that market-based capitalism is a faith system. While many scholars agree with his claim that 'without certain theological assumptions, some of the most important conclusions of economic theory could not sustained' (2001: 65), many also use this conclusion to debunk economics rather than to exalt it.

David R. Loy gives the most compelling of the many critiques of the 'economics as religion' argument. He does this through looking more deeply into the metaphor. Loy warns us 'Nelson … could be said to have overlooked the market religion's sacrificial aspects of worsening global poverty and environmental degradation' (cited in Foltz 2007: 146). We might imagine that this situation arises somehow accidentally, an unanticipated outcome; however, closer inspection reveals a much more sinister picture.

In July 2010, in *The Independent*, journalist Johann Hari (2010) observed that in 2007 staple food prices skyrocketed and another 200 million people – mostly children – were condemned to hunger, malnutrition or starvation. There were riots in more than thirty countries. However, these price rises weren't a result of a fall in the supply of these foods. Wheat production actually rose across the globe that year. Then, in spring 2008, prices just as mysteriously fell back to their previous level. Hari cites Jean Ziegler, the United Nations Special Rapporteur on the Right to Food, who called this 'a silent mass murder', entirely due to 'man-made actions'.

Here's why it happened. Throughout the twentieth century farmers have had mechanisms, futures markets, that allow them to insure themselves against crop failure or the collapse of prices. Throughout the 1990s Wall Street traders lobbied for the abolition of the regulations that had hitherto restricted these mechanisms so that only those directly involved in food production could use them and suddenly these contracts were transformed into food speculation derivatives that could be traded without reference to the foodstuffs themselves. Previously the economic system of supply and demand ensured that 'only' a billion people went to bed hungry. After this deregulation, the cost of food was determined by the value of speculative food contracts not by the availability of the food itself, and millions more were unable to afford to eat sufficiently.

Asked if this was a bubble that's being deliberately created, Jayati Ghosh, Professor of Economics at the Jawaharal Nehru University in New Delhi, points out that the price of crops that are **not** traded on the futures markets, like millet and cassava rose very little in 2007. She concludes that speculation was the main cause of the rise. John Lanchester, also quoted by Hari, argues that

> finance, like other forms of human behaviour, underwent a change in the 20th century, a shift equivalent to the emergence of modernism in the arts – a break with common sense, a turn towards self-referentiality and abstraction and notions that couldn't be explained in workaday English.

The result was, Hari argues,

> In 2006, financial speculators like Goldmans pulled out of the collapsing US real estate market. They reckoned food prices would stay steady or rise while the rest of the economy tanked, so they switched their funds there. Suddenly, the world's frightened investors stampeded onto this ground.

While the supply and demand of food stayed pretty much the same, the supply and demand of derivatives based on food massively rose. This meant the all-rolled-into-one price shot up, and greatly increased starvation arose. The bubble only burst in March 2008 when the situation got so bad in the US that speculators had to slash their spending to cover their losses at home. Hari (2010) concludes,

> The world's wealthiest speculators set up a casino where the chips were the stomachs of hundreds of millions of innocent people. They gambled on increasing starvation, and won. Their Wasteland moment created a real wasteland. What does it say about our political and economic system that we can so casually inflict so much pain?

In 2009, in order to prevent recession (the failure of the economic deity to grow), many governments, around the world, initiated a 'stimulus': a sacrifice to economic growth. In Australia this was done by giving most taxpayers $900 in the hope that if enough people spent this money the god of economic growth would be mollified and economic growth would recommence. This was a response to the obvious: when the god gets wrathful, He stops growing, we get depressed, people lose their jobs and the economic consequences of this emotional instability multiply.

It seems to me that people bow to the ideological reign of economics today in the same sort of way as they bowed to the reign of inquisitorial Christianity in the Middle Ages. In those days a punitive, feudal Christianity permeated all aspects of life, and individuals were compelled to demonstrate their piety. That fervor looks bizarre to us today. Perhaps one day our own fealty to the hegemony of economics will look just as bizarre.

Speaking truth to power

From a deep ecology perspective, the corruption of soul, society and soil by contemporary economic thinking is pernicious and possibly terminal. But what are we to do? It's not much use pointing fingers or claiming the high moral ground: nary a person is exempt from this religion nowadays for – to stick with the metaphor – even if we are not active in the congregation, all of us are members of the sect. All of us are fouling the Earth. Yes, some have their hands on bigger triggers but we are all caught in this psychotic dance.

For this reason I would like to conclude by exploring some avenues whereby we might respond to and/or resist this overwhelming economic determination of our experience.

I believe that we need activism, not solely of the traditional political kind, but activism that incorporates a philosophical or 'spiritual' movement to unmask these expansive powers. This should take the form of a 'hearts and minds' campaign to which we apply the same fervour that we bring to our campaigns to protect a special place or an endangered species. The special place will be the Earth and the endangered species will be biodiversity, for no place or species is safe from the consequences of the policies of the economic theocracy.

Bringing the force of the theocratic metaphor to this discussion, we must throw the moneylenders out of the Temple of the Immaculate Biosphere. We must defrock economics and strip it of its plausibility, rescind its Nobel Prize, humiliate it and provoke laughter at the posturing of the naked emperor and his obsequious courtiers.

This sort of social–ecological satire is something I have been engaged in for more than twenty years. In part court jester, in part bittersweet clown, I have mixed relentless critique with comedy to draw attention to structural hypocrisies that condemn our species to an all too rapid decline. I am able to do this because I invest as little as possible in the financial system. I harvest my idealism and tithe my entertainment. This makes it easier for me to identify with and celebrate the ridiculous within society and myself. This is not something I do alone. I often find myself, unexpectedly, in remarkable company. For example Sir David Attenborough, David Suzuki, Olivia Newton John and Jack Thompson all gave their time free of charge to appear in my 2003 forest campaigning film *On The Brink*.[1] This campaigning of mine has been acknowledged to the extent that, through some unusual coalescence of forces, the federal government of Australia saw fit to celebrate my efforts with an Order of Australia Medal.

Ironic laughter please!

We need to highlight the fact that when we see economics as a religion, then advertising becomes religious education. Ellul calls advertising 'the liturgy and the psalmody of the consumer religion' (1975: 3). I believe that a critique of advertising is a strategic place to begin to undermine the religion.

This work, of exposing and deconstructing advertising, is already underway in some places. Adbusters magazine (www.adbusters.org) uses satire to strip the emperor of the veils of illusion behind which he hides; the 'Story of Stuff' (www.storyofstuff.com) offers a popular and accessible online analysis while Reverend Billy (www.revbilly.com) of the 'Church of Life After Shopping' wittily thumbs his nose at the false god. These critiques are examples of a rapidly emerging suspicion around the prevalence of public relations and spin in the corporate sector and in government. Nevertheless, the machine that requires and maintains advertising is extremely powerful. More laughter is required!

I believe we need to build these beginnings into a movement that redefines what it means to be successful, what it means to be abundant. This movement needs to be based in a depth of understanding of ecological relationships.

We need to emphasize the value of the respect that stems from a profound understanding of the connection between humans and nature and between humans themselves. Respect is subscribed in laws as rights – the requirement that the rights of others are respected. These need to include the rights of future generations and the rights of other life forms. Arne Naess called on us to 'heal our relations with the widest communities, that of all living beings' (cited in Seed et al. 1988: 29). He, like I, regards human rights as inalienable and are not simply that which is granted by a state. They are rights that we hold against the interests and actions of a government/state. The same goes for rights in nature.

More laughter, is needed, so much more!

Fueled by advertising, we dig up the Earth up and chop it down its fruit to make the goods we presume will satisfy our soul.

More laughter, more, more, more.

Like any addict we must ask: 'What is the real underlying problem? What is it we are not facing up to and avoiding by our consumption habits?'

While the false religion of economics needs to be demolished, the religion of the sacred cosmos needs adherents. The mystique of the Earth is a primary requirement for the establishment of a viable rapport between humans and the Earth. Here we may find guidance in the work of Thomas Berry (1988). Berry was a Catholic priest who, during the middle period of his life as one, he turned from a theologian to a 'geologian': someone who found his source of spiritual inspiration and nourishment in the Earth itself. Berry influenced and was influenced by the deep ecology movement. A student of Teilhard de Chardin, Berry proposed a deep understanding of the history and functioning of the evolving universe as a necessary inspiration and guide for our own effective functioning as individuals and as a species. He writes:

> If the dynamics of the universe from the beginning shaped the course of the heavens, lighted the sun and formed the Earth, if this same dynamism brought forth the continents and seas and atmosphere, if it awakened life in the primordial cell and then brought into being the unnumbered variety of living beings, and finally brought us into being and guided us safely through the turbulent centuries, there is reason to believe that this same guiding process is precisely what has awakened in us our present understanding of ourselves

> and our relation to this stupendous process. Sensitized to such guidance from the very structure and functioning of the universe, we can have confidence in the future that awaits the human venture.
>
> (Berry 1988: 137)

Acknowledgements

I am deeply indebted to the work of Richard Foltz (2007). It was discussions with Richard starting in 2003 that sparked my research into this topic.

Note

1 www.youtube.com/user/rainforestinfo?feature=mhum#p/u/5/gnnyaDZ3bck, accessed 10 July 2010.

References

Berry, Thomas (1988) *The Dream of the Earth*, San Francisco, CA: Sierra Club Books.

Brown, Norman O. (1959) *Life Against Death: The Psychoanalytical Meaning of History*, Middletown, CT: Wesleyan University Press.

Cox, Harvey (1999) 'The market as God: Living in the new dispensation', *Atlantic Monthly*, March, 18–23.

Ellul, Jacques (1975) *The New Demons* (trans. C. Edward Hopkin), New York: Seabury.

Foltz, Richard (2007) 'Religion of the market: Reflections on a decade of discussion', *Worldviews*, 11: 135–54.

Hari, J. (2010) 'How Goldman gambled on starvation', *The Independent* 10 July, http://www.independent.co.uk/opinion/commentators/johann-hari/johann-hari-how-goldman-gambled-on-starvation-2016088.html, accessed 10 July 2010.

Næss, Arne (1973) 'The shallow and the deep, long-range ecology movement', *Inquiry*, 16: 95–100.

Nelson, Robert (2001) *Economics as Religion*, University Park, PA: Penn State Press.

Reverend Billy at http://www.revbilly.com/accessed 23rd March 2011.

Seed, John, Fleming, Pat, Macy, Joanna and Naess, Arne (1988) *Thinking Like a Mountain: Towards a Council of All Beings*, Philadelphia, PA: New Society Publishers.

'Story of Stuff' http://www.storyofstuff.com./accessed 23rd March 2011.

26

A DRAMA ECOLOGY OF CULTURE

Ben-Zion Weiss

> The planetary ecological crisis allows us to see the nature of a planetary ecology for the first time. If we can begin to understand the pattern that connects … nature to culture, we have the possibility of becoming alive in vitally more imaginative ways …
>
> (Thompson, 2001: 17)

This chapter explores a drama ecology of culture in order to understand the pattern that connects 'nature to culture' and how this can make us become more alive. It is informed by my academic research in drama and theatre studies as well as my experience in theatre and in drama education over the last three decades.[1] It addresses vital issues in our contemporary world in relation to the planetary ecological crisis. A crisis I am framing as a crisis in our cultural ecology. My writing in this chapter is framed within the discipline of social ecology, which includes perspectives drawn from personal, social, ecological and spiritual domains. I begin with the personal.

Introducing myself as researcher

> All scholarship is disguised autobiography …
>
> (Thompson, 2001: 36)

My passion for drama education grew out of my experience in theatre as well as my experience as a youth worker and as a teacher. After some years of acting, directing, writing and technical production in theatre, film and television, I realized that as much as I loved the creativity and the opportunity for expressing myself, this was not going to be an easy way to make a living. I was facing the classical dilemma of the artist and social activist in the contemporary world of commerce, science, technology and industry. How was I going to survive financially by making radical theatre or films?

After reflection, soul searching and some great conversations with friends and mentors, I decided to be a drama teacher. Then I could get paid for my art as well as influence the society I lived in through educating the next generation. I returned to university and completed my Arts degree in Drama and a Diploma in Education in Drama and ESL (English as a Second Language), for secondary schools.

After graduating, I worked as an actor/teacher for a theatre-in-education group in south-western Sydney. We toured schools running Playback Theatre workshops and getting the students to perform for their peers. The power and effectiveness of this model was to repeat itself many times in my career as a drama educator. It led to a Masters in Theatre Studies, researching Playback Theatre and finally a PhD in Social Ecology focusing on using drama education for anti-racism, at the University of Western Sydney (UWS).

I chose to research anti-racism for my PhD using drama education because it involved experiential learning. It is a process that can engage deep community attitudes. As a Jewish person growing up in Australia, I experienced some racism. However, I discovered indigenous people, certain migrants and refugees, experienced considerably more racism. I was aware of the profound ecological knowledge that indigenous cultures have with their deep connection to nature. Racism towards them prevented this relationship to nature being made available to the wider society. My thesis was called *Challenging Understandings of Racism through Drama Education Praxis: Steps to an Ecology of Culture* (2007). Next, I define my terms before I describe the research itself.

Defining key terms: Ecology, culture and drama

Ecology is a relatively new field of scientific endeavour. It emerged from the work of the German biologist Ernst Haeckel who wrote in 1866 'each individual living organism is the product of cooperation between its environment and the body it has inherited' (Allaby 1986: 11). 'Culture' is a very difficult word to define. The cultural theorist, Avruch, came up with 164 definitions (2002: 109). He proposes Theodore Schwartz's:

> Culture consists of the derivatives of experience, more or less organised, learned or created by the individuals of a population, including those images or encodements and their interpretations (meanings) transmitted from past generations, from contemporaries, or formed by the individuals themselves.
> (2002: 17)

These definitions clarified my understandings about drama education.

A primary form of encodement of culture is drama. In the words of the drama educator, Richard Courtney (1980: 1): 'Life is a drama. Always we act roles. Our clothes are our costumes, and our setting is the space in which we act.' What better way to understand the pattern that connects culture and nature or the processes of life than through drama?

Drama involves actors playing roles. In my PhD, I argued that our cultural identity is a role we learn to play as part of learning to live in our culture. I deconstructed theories of race, and examined how the development of contemporary racism related to the Holocaust; power issues in relation to dominant cultures, colonialism, indigenous cultures; and the idea of culture. I address Australian multiculturalism and how an ecology of culture provided a possible next step in dealing creatively with cultural diversity.

The research

My research with youth and youth workers in training, involved a hybrid methodology: an intuitive narrative inquiry; and three stages of a creative action research process. Intuitive narrative inquiry allowed me to pursue the story of the research in an intuitive way, as only the first stage of the action research was planned. Stages two and three emerged through my work as a drama educator and a social activist for anti-racism.

Over a number of years of consulting, training and teaching in the area of non-violence, creative conflict resolution, community drama, cross-cultural communication and anti-racism, I developed a workshop for training youth workers in anti-racism. I called this 'Cooling the mix: Dealing with youth racism creatively'. This was the focus of the first stage of the research with youth work students. The second and third stages were located at an inner city high school with a high proportion of Aboriginal and Pacific Islander students.

One of the challenges of racism is that people express racist views or act in racist ways quite unconsciously. Reframing racism as cross-cultural conflict allowed me to use strategies from peace and conflict studies that drew on theories of non-violence training, conflict resolution, peace activism and education. A prime example of this was reported in one of the post-workshop focus groups, when a young non-indigenous woman returned to her parents' house in the suburbs and was confronted by her brother's work colleague, who expressed racist views towards Aboriginal people. Instead of her usual angry reaction, she engaged the man in a conversation about the issue. She considered this more effective than 'just having a stand up fight with him'. From similar other observations by participants and my own observations that racism could be reframed as cross-cultural conflict, I concluded that my research was effective.

Through my work with indigenous people I realized that indigenous cultures did not separate ecology from culture. This was also true in Ancient Greek culture, one of the origins of contemporary Western theatre. The Ancient Greeks used theatre as ceremony to celebrate the God Dionysus. Culture and ecology were interwoven in the ancient story of Dionysus, God of theatre, wine, wilderness and the ecstatic experience, all of which involve the process of transformation. Through such transformational processes, cultural conflicts like racism can be transformed through drama education into intercultural communication and education.

In stage two of the creative action research, we produced the radio show called *Turn it up on Anti-Racism*, which was broadcast on a number of Sydney community radio stations. Improvisation workshops based on the participant's experience of racism, were developed into the radio show. It was a creative process that engaged high school students, youth workers, teachers and Aboriginal elders. Radio plays, role plays and rap songs were written and performed, interviews were conducted, stories were told, conversations were generated, all out of the enviro-scape of an inner city school, a cultural camp on the Central Coast of New South Wales (NSW) and the inner city suburbs where the people lived their lives.

I use the word *enviro-scape* to bring together the scientific term *environment* and the artistic word *landscape.* Artists depict their environment as landscapes making culture, and scientists study landscapes as environments to understand the ecology of a living species. In this term artistic and scientific understandings are brought together. Creativity is embedded in the enviro-scape of the creative person. A grounded example of the creative process is an ecology of culture, where culture grows out of the enviro-scape. I argue that a drama ecology of culture can help us to 'begin to understand the pattern that connects … nature to culture' (Thompson 2001: 17).

In Australia, the British colonizers imposed their culture on the indigenous inhabitants, as well as on the enviro-scape itself, with devastating effects on both. This imposition of British culture continued right up to the early 1970s when multiculturalism was adopted as official policy, opening the door for a more inclusive contemporary Australian culture to emerge. A key to liberating Australian culture from British colonial culture is spontaneity. Spontaneity grows out of the present moment and allows us to transform the past through our creativity, rather than reproducing it as victims of history.

A 'student-centred' approach to learning was fundamental to dealing creatively with the indigenous students' conflict situations in their experience in DRACON,[2] the third stage of my creative action research. DRACON is a hybrid word formed from the words 'drama' and 'conflict'. It involves students acting out conflict situations from their lives as a way of deepening their understanding of conflict situations and how to deal with them more creatively. They then peer teach the process to younger students. DRACON uses a modified form of Forum Theatre, as developed from Boal's (1979, 1999) work, and also involves elements of Playback Theatre, which has many similarities to Forum Theatre.

Playback and Forum Theatre involve the use of personal stories from participants, which adds a sacred dimension as people share from an intimate space. These forms allow stories of audience members or workshop participants to provide the content of the drama demonstrating Boal's insight that 'Observing itself, the human being perceives what it is, discovers what it is not and imagines what it could become' (1999: 13).

The theory of drama education

Drama education addresses the 'empty formalism' that radical educational thinker Paolo Freire identifies in education:

> Creativity does not develop within an empty formalism, but within the 'praxis' of human beings with each other in the world. In this 'praxis', action and reflection constantly and mutually illuminate each other. Its practice, which involves a theory from which it is inseparable, also implies the attitude of someone seeking knowledge, and not someone passively receiving it. Thus, when education is not truly a gnosiological condition, it diminishes it to mere verbalism which, because it frustrates, is not inconsequential.
>
> (1975: 149)

This mere 'verbalism' in high school education is something I observed as a classroom teacher. Much schoolwork given in classrooms that I supported as an ESL or literacy teacher, did not engage students fully. Educational drama activities can be a way to engage students and to encourage greater participation in their learning process.

This has resonances with the drama education movement, which also identifies the lack of 'affective orientation' (Johnstone 1982: 17) as a major limit of Western education. They see drama education as providing the form to remedy this limitation. One of the major theorists in this movement, Gavin Bolton, argues:

> We have assumed that pedagogy is the training of children in the neutral observation of objective facts. Teachers have often paid lip service to, or ignored the affective orientation or, equally mistakenly, have assumed that such an orientation means free expression rather than understanding. In my view, both orientations are directed towards the development of concepts and we have a responsibility to make both modes available to the children we teach. Not only do we neglect the affective orientation but often we train children to despise it.
>
> (1979: 39)

For Bolton (1979) it is this affective dimension that enables 'drama for understanding' to develop as a methodology. It encourages the development of emotional intelligence (Goleman 1996), which is empowering for members of marginalized groups.

A powerful form of drama education is Playback Theatre. In playback, the dramatic action is already a 'reflection' of a story, told by a participant, who can then 'reflect' on that 'action' afresh, as can other people present. As such, it may be regarded as a particular case of Freire's *praxis* idea, as well as Bolton's 'drama for understanding'.

Playback affirms the experience of individual audience members, who are initially 'warmed up' to telling longer stories by the 'conductor', who invites members of the audience to share some 'special' moments and stories from their life. The form was created by American psycho-dramatist, Jonathan Fox (see Salas 1983). Playback has relevance to the formation of identity for young people in their growing up process, especially when their stories are different from those they receive in mainstream media. Young people from diverse cultural backgrounds often complain of not seeing images of people like themselves in Australian public media,[3] which leads to a sense of not belonging to what they perceive of as Australian culture.

If we are not in touch with ourselves, how can we hope to be in touch with our full human potential? If we do not hear and see our own stories in public, how can we get a sense of who we really are and what our cultural identity is?

Augusto Boal and Forum Theatre

O'Toole and Burton (2001) in their theoretical material supporting DRACON observe that:

> An equally influential worldwide movement of drama education: *Theatre of the Oppressed* (TO) has grown up contemporaneously with the *Drama in Education* (DE) movement – in some places until recently almost independently. The Brazilian theatre director Augusto Boal founded this. Directly inspired by Paolo Freire's *Pedagogy of the Oppressed*, Boal set out to develop a community theatre practice that has become, in effect a pedagogy, whose original aim was to help the poor and disenfranchised to liberate themselves by revealing the nature and conquerability of those things that oppressed them.
>
> (Module 6.2: 4)

Boal's work provides another angle on the process of developing a sense of identity and a consciousness of self in community, which has powerful possibilities when applied to issues involving conflict between different groups in a community, in particular cross-cultural conflict. In NSW, the Multicultural Programs Unit (MPU) implemented DRACON as part of the Whole School Anti-Racism Project (WSARP).

The success of the anti-racism radio show drew the attention of the MPU to the work we were doing and resulted in the school being included in a major research project by Griffith University's Applied Theatre Research Unit on DRACON in NSW schools as part of WSARP. DRACON had been implemented in high schools in the Brisbane area and in regional NSW, where it was shown to be effective in dealing with incidents of racism. It empowered students who were the victims of racism. I had observed how these students are sometimes identified as having 'behaviour issues' as they can act out their anger and frustration in school contexts. Drama education can provide such students with other ways of communicating their experience. Such students were found to act as leaders in the program as part of the peer teaching component, as when three Koori – Aboriginal – year 11 students peer taught a class of sixteen year 5/6 students at one of the feeder primary schools.

Some concluding comments

Drama as a way of learning can be regarded as an example of an ecology of culture. Ecology itself involves systemic interactions between different actors in the ecosystem, which in this case are the different cultures that make up the ecology of culture. This is located in the field of social ecology, which integrates personal, social, ecological and spiritual perspectives.

By developing our creativity through improvisational forms of drama that we create out of our life stories and experiences, we gain insights and understandings of the choices we make through actions, words, thoughts and feelings and especially when we give ourselves the opportunity to reflect on those experiences. This helps our understanding of the complex processes in which we engage, as we play out the diversity of roles into which we have been acculturated by the culture we have inherited. In the highly diverse society that contemporary urban Australians face on a daily basis, there are many social roles we need to play, in our day-to-day life.

In any human life, conflicts arise through misunderstandings, competing interests or needs – to name just some of the possible causes of conflict. These are likely to be aggravated by cultural differences. Conflicts may escalate or de-escalate, depending on the kinds of decisions made. My argument is that the more conscious we are of this complexity of roles, the more creative we can be in our decisions.

High school students and youth, through their experience of drama education, can become conscious of the roles they are playing. Therefore they have a greater range of choices in the decisions they make. To deepen understandings of these roles, and of the conflicts that arise in cross-cultural conflicts like racism, requires critical reflection. Such reflection can develop more understanding of an ecology of culture, which in turn can lead to addressing the planetary ecological crisis, mentioned in the quote at the beginning of this chapter.

For myself, this research work has opened up new areas of teaching and workshop facilitation. It has involved teaching social ecology subjects at the University of Western Sydney and working as a consultant for the Multicultural Unit of the NSW Department of Education and Training. The latter has involved training teachers in over 100 schools in Sydney and regional NSW in the implementation of the Cooling Conflicts process, which has proven effective at addressing issues of racism and bullying in schools (O'Toole et al. 2005). In the undergraduate social ecology subjects, the majority of students are teachers in training. The quality of work produced by these students, their enthusiasm for these units and their end of unit feedback all indicated that their experience of learning in these areas was positive and beneficial to the learning process.

Notes

1 This chapter is adapted from a paper given at the 2006 Drama Australia Conference at NIDA: 'Turning the tides', which was based on my masters and my doctoral research in drama education.

2 DRACON (now called Cooling Conflicts) is a process that originated in Sweden, in the Peace and Development Research Institute of Gothenburg University, and was brought to Australia by Professors John O'Toole and Bruce Burton (2001) from the Griffith University's Centre of Applied Theatre Research. The program had been trialled and deployed, initially in high schools in the Brisbane area and in NSW. Cooling Conflicts has now been implemented in over 100 schools in NSW.

3 Australian born Chinese politician, youth advocate and lawyer, Jason Yatsun-Li spoke of this at *The World Peace Summit* at The Great Hall at Sydney University, 1 March 2003.

References

Allaby, M. (1986) *Ecology Facts*, London: Hamlyn.

Avruch, K. (2002) *Culture and Conflict Resolution*, Washington, DC: US Institute of Peace Press.

Boal, A. (1979) *Theatre of the Oppressed*, London: Pluto.

Boal, A. (1999) *The Rainbow of Desire, The Boal Method of Theatre and Therapy*, London: Routledge.

Bolton, G. (1979) *Towards a Theory of Drama in Education*, London: Longman.

Courtney R. (1980) *The Dramatic Curriculum*, London: Heinemann.

Freire, P. (1975) *Pedagogy of the Oppressed*, Harmondsworth: Penguin.

Goleman, D. (1996) *Emotional Intelligence*, London: Bloomsbury.

Johnstone, K. (1982) *Impro: Improvisation and the Theatre*, London: Faber.

O'Toole, J. and Burton, B. (2001) *The DRACON Conflict Management Program Manual*, Queensland: Griffith University.

O'Toole, J., Burton, B. and Plunkett, A. (2005) *Cooling Conflict*, Sydney: Pearson.

Salas, Jo (1983) 'Culture and community: Playback Theatre', *TDR*, 27(2) (T98), Summer.

Thompson, W.I. (2001) *Transforming History, a Curriculum for Cultural Evolution*, Great Barrington, MA: Lindisfarne.

Weiss, B.-Z. (2007) *Challenging Understandings of Racism, through a Drama Education Praxis: Steps to an Ecology of Culture*, PhD thesis, University of Western Sydney.

Further reading

Bateson, G. (1972) *Steps to an Ecology of Mind*, NY: Ballantine.

Boal, A. (1996) 'Politics, education and change', in O'Toole, J. and Donelan, K. (eds) *Drama, Culture and Empowerment, The Idea Dialogues*, Brisbane: Idea.

O'Toole, J. and Burton, B. (2002) 'Article No.1, Cycles of harmony: Action research into the effects of drama on conflict management in schools', *Applied Theatre Researcher*, ISSN 1443-1726 Number 3.

O'Toole, J. and Donelan, K. (eds) (1996) *Drama, Culture and Empowerment, The IDEA Dialogues*, Brisbane: IDEA.

Slade, P. (1975) *An Introduction to Child Drama*, London: Hodder & Stoughton.

Soyinka, W. (1973) *The Bacchae of Euripides*, Oxford: Oxford University Press.

Weiss, B.-Z. (1986) *Explorations of Improvisational Drama – An M.A. Project*, Master's thesis, University of NSW, School of Theatre Studies.

Wright, D. (2000) 'Drama education: A "self-organising system" in pursuit of learning', *Research in Drama Education*, 5(1): 24–31.

27

MAPPING MACHANS BEACH

Meandering in place (a beginning)

Susanne Gannon

Some terrain must be mapped at ground level, step by steady step. Poetic form also slows the pace of language, brings the particular into sight, creates rhythm and space, demands precision and love. Poetic mapping is one of many possible strategies for exploring one's sense of place within social ecology. Personal explorations of place and space through all sorts of media, through narrative, through dreams, through memories can all be facets of the sort of 'place responsiveness' and 'deep engagement' with personally significant places that Cameron advocates as an essential and initial aspect of environmentally sustainable ethical practice (2003). Cameron's approach to educating for place responsiveness couples critical thinking with this deep affective engagement for powerful pedagogical purposes (2003: 105). This chapter demonstrates my own idiosyncratic response to the call to engage with my particular place and space, through my particular preferred medium of language. It is also informed by Massey's understanding of space as 'a plurality of trajectories, a simultaneity of stories-so-far', each told from a particular and only ever momentary angle of vision (2005: 12). Part of this chapter presents my own text of place, from a particular angle of vision, my own slow meander around memories of place, but I also want to make an argument for opening modes of academic response in social ecology to encompass poetic texts.

In the first section of this chapter, my poem takes a walk through a coastal town in the northern tropics, while the second section sketches out a poetic method that might be taken up to craft responses in social ecology inquiry. The poem in this chapter forms a material space where soils, vegetation, birds, animals, light, landscape, people and their habitats coexist on the same plane of being. This intimate landscape is thickened by memory. The poem begins with a topographical sketch, then sets off on a morning walk around the block. With each step the poet takes – along the water, around corners and past houses – details are documented and stories of that place arise, intersect and fall away to be replaced by other unresolvable stories. Each step in the present opens steps into the past. Throughout the poem, the conventions of narrative genre are undone as temporality is disjointed and events and people are elusive and impressionistic. This poetic map of memory trails towards an end that is only ever a beginning, again and again. With a beginning, an end, a series of events in time and place, and a cast of characters – human and non-human – the poem can be understood as a sort of narrative told through the prism provided by the voice of the narrator-poet. However, rather than a tidy seamlessness, this is a narrative produced by (and producing) 'multiple, disunified subjectivities' rather than 'singular agentic storytellers' (Squire, Andrews and Tamboukou 2008:3). Through this poem, the particular neighbourhood that is mapped, step by step, becomes a central protagonist of this complex narrative of place.

Mapping Machans Beach

1

Bound by water, Barr Creek blocks the way north
except when tide fall and sand drift open a path
to Holloways Beach and the next demographic.
Huge houses fringed by fallen palms, tilting
closer to the Coral Sea with each king tide.

Some Sundays, I'd miss the low tide
carry my clothes on my head, wade back
chest high in water, thinking of young bulls
floating out to sea and of crocodiles
hungry and heading for home.

To the south, the dark waters of Redden Creek
flooded mangroves, silence broken by *tt–tt–tt–tt*
of tinnies at dawn, men leaning on the bridge,
cast-nets like orchids flowering and falling
rising wet and silver with fish.

Further south, the black slash of the Barron River
meets the sea. Across the water, sheets of tin glitter
from camps at Dungara, the new-old name
for the far shore, home for fishermen and ghosts.

There are low tide runnels and peninsulas
of toe sucking mud, a grandmother's ashes,
fragments of burnt bone, sand, seaweed and shells
beached along the high water mark.

East is water all the way to Vanuatu, drowned ridges
of reef, the grief of excised borders. Granite boulders
mark the land's end, slapped together with
home-mixed concrete, stubbled with debris,
driftwood, tumbled terracotta bricks.

At low tide, crabs scuttle into shade. Locals promenade
on bitumen in sarongs and thongs, stubbies and shorts.
Morning and evening shifts for the workers, stray
fisherfolk and lost tourists looking for a beach.

Westward, past streets of fibro houses, a patchwork
of cane stubble, green blades and feathered flowers
taller than a man. The creek curves by the road, marking
one, two, three metres above sea level.

Remember post-cyclone weeks when only the post office
dinghy could get us and the beer in and out, helicopter
drops on the radar field of milk and bread and methadone.
Storm surge maps show all of it under, all of it gone.

2

Half light and I rise to the *owk owk* of friarbirds. Turning
up Marshall St, I slam the lattice gate behind me,
damp wood distended by old rain. Always looping
around the back to begin again, saving the front
for a southward tilt towards the early sun.

Bitumen turns to sand and I track behind houses.
On my left a broad expanse of kikuyu, the bush beyond
where kids and kerosene collided one night in a carnival
of sirens and firetrucks and crackling spitting flames.

Step on the grass and the skirr of rainbow bee-eaters
rises to meet you, wings blurring yellow, green and black.
Ragged orchards of pink-tipped mangoes, bananas,
beach almonds, stray vehicles, swings, scruffy backsides
of houses to the front, a verandah that was almost mine.

I skittered away from home then, spooked by phone calls
and a season of break and enters, stepped in to the rooms
of a writer split between New York and the islands,
took comfort in her two quiet rooms, the sleeping one
and the writing one, and her long low view to the hills.

Following the sand sweep around to the ocean, thongs
looped in my fingers, I glimpse the dreadlocked man
from Fortaleza I met here, his slow arabesque
of dark legs and hair spinning against the blue sky,
capoeira on the beach at dawn. *Bom dia ... Moito bom.*

I step up the new metal grids, missing the old angles,
higgledy piggledy blocks of warm stone. Here
once, when the tide was high and the light low,
we floated flowers for a neighbour we didn't help
who cared too much for love and followed it too far.

3

On the Esplanade, I come up by the stinger vinegar
and peel-off dog-shit bags. A gull sits on the
No parking, No camping sign. A Wicked camper stirs
in the morning heat, groans open, unfurls
long blonde limbs in fisherman's pants.

Progress here is a straggly vacant lot and our struggle
for 'Red Dirt Beach', a trucking magnate dreaming
of shiny townhouses and ocean views, and our ragtag
rabble of posters and protests and EPA regulations.

The palms along the waterfront are bereft of coconuts
Two of the old melaleucas gone now from the house I shared,
so too is the friend who might have died anyday,
who tangoes with a new heart half a continent away.

Then the home of everyone's *querida,* gone to Timor Leste
with the weaving women. I remember a *telenovela* Easter
when she fell through the floor and rose up laughing.
The hammock is furled under the eaves now and
a satellite dish on the roof signals the changes.

There's the blue and orange doctor's house, long grassed
and littered with sturdy children and plastic climbing frames.
Further on, the same blue pram as yesterday sits by the road,
damp from overnight rain. There is a boom of babies
named for stars or minerals or movements of the earth.

Here's the house with red blinds like eyelids. The son-in-law
who lived beside me one street back, the calls to cops and threats
and screams, his wife a pale blonde wearing plaster,
their quivering child, the over-the-fence escape route home
to mum and dad. That boy would be at school by now.

4

My ankle twitches at the paperbark roundabout where
Arnold St hits the beach. The bitumen is smooth now,
the council wary of the just a couple of glasses
after dinner and a walk around the block stumble after dark,
though that was not the only time and not the only damage.

To my right is the wooden louvred house famous
for parties of tumblers and poets, pineapple risotto and PhDs.
A stranger on an esky is the just left yesterday
husband of my best friend when I was seven,
three states away and in the next suburb.

We have nothing in common, but I file her tales of cold
clothed showers in the convent next to memories of
kneeling for the rosary before tea in her house
and how the rooms went on for ever and now she skippers
her own big boat on the Coral Sea.

I could go on, and do on other mornings, past houses
and people I've loved, places where waves broke
over me, and showers of comets fell from the sky.
But, this day, knowing that I'm losing my place,
I turn back to the pink house with the swollen gate.

This poem, this beginning of a poem – twenty-five stanzas long, twenty-five years after I first moved to north Queensland – ends here, for the moment. It is the beginning of an attempt to map a place, to map memories of a place, to map self in a place peopled in the past and present, a place replete with non-human inhabitants, a place struggling with coastal erosion, development pressures and fragile ecologies. No matter how steady and careful the steps might be in a walk, or in a poem, no one – not even the poet herself – can stand still long enough for this sort of map work. The poem began during a visit back to the small suburb where I had lived for most of my years up north. I had been loaned a house by a friend for a fortnight, just up the street from the last house I had lived in there, just around the corner from other houses I had known. Just a block from the beach in the front, the house backed on to what we called the 'radar field', a navigation aid for pilots heading for the airport nearby, but from our angle close to the ground it was a vast swathe of open space saved for the moment from development. In that house for that fortnight, looking at familiar landscapes, and on walks where memories rose up towards me, tempered by ambivalence at having left my 'home' five years earlier, I tumbled into a textual reverie where I tried to bring all this together.

Some thoughts on poetic method and social ecology

To my knowledge poetry has always been a feature of the social ecology programs at the University of Western Sydney. Usually one evening is set aside in each of the intensives for 'poetry and performance' where participants read poetry or other texts by others or themselves. They may sing, tell a story, show a film, or lead a dance or a meditation. Although these practices might seem to some to be peripheral to the serious business of the day, they are at the heart of the practices of community building that sustain people through academic work. They are also at the heart of the social ecology curriculum, where creative engagements through art, theatre and critical autobiography are central components of a co-created 'ecological learning web' (Camden-Pratt 2008). However, in the argument I want to make in this section of the chapter, I want to tease poetry apart from other modes of creative inquiry and representation and consider – through the poem in this chapter – its particular affordances for social ecology research.

Poetry can be understood as a legitimate form of inquiry in its own right, requiring deep listening, a recognition of the inevitable inadequacy of language and its ethical implications, and the necessity that research must always be complex and difficult work (Richardson 1997; Neilsen 2004; Prendergast 2009). In its attention to the rhythms of breath, and space, to embodied and affective experience, to the connectedness of human and non-human, to spirituality and environment, poetry honours this complexity. Poetry can enable the holistic way of working with materials that acknowledges that personal, social, environmental and spiritual modes of being and knowing are interdependent (Hill 2003; Hill et al. 2004; Camden-Pratt 2008). Poetry brings together 'embodying, emotioning, expressing' (Wright 2005).

The poem in this chapter is part of a larger investigation into place and its representation in text. As well as the work of social ecologists, I have been inspired by Foucault who suggested that 'our experience of the world is less that of a long life developing through time than that of a network that connects points and intersects with its own skein' (1986: 22). His spaces of 'heterotopia' are simultaneously real and unreal, represented, contested and inverted like a mirror that makes

> the place that I occupy at the moment when I look in the glass at once absolutely real, connected with all the space that surrounds in, and absolutely unreal, since in order to be perceived it has to pass through this virtual point which is over there.
>
> (1986: 24)

I have been inspired by the 'poetics of space' of Bachelard (1964) to conduct a poetic 'topoanalysis' of the intimate sites of my life, hoping that my memories of place will echo and reverberate with others who might read them (Gannon 2009). This sort of inquiry, in social ecology and in other fields where a heightened aesthetic and ethical attention to language is brought to bear on the text, can be understood within the field of arts-based research and practice (Cahnmann-Taylor and Siegesmund 2008). Such work endeavours to move readers simultaneously on emotional, affective, intellectual, and aesthetic levels. However, unlike visual modes of arts-based inquiry, poetry has only words and sounds and patterns on a page with which to make moves on its readers.

The craft of poetry – rhythm, prosody, form, imagery, metaphor just to start with – must be attended to in this sort of writing. It is not easier to write poetry than prose (though prose may also carry the potency of poetic language). As in any form of writing in (or outside) academia, some knowledge of the genre of text is important. It opens possibilities for revision and reversals, for crafting the work beyond its first raw rough emergence. Read poetry if you want to write poetry. Learn as much as you can about what poetry can do, and take this as far as you can. Poems often find their own form, but this poem in this chapter is loosely shaped around a 'retrospective-prospective structure', a temporal oscillation between past experiences, feelings or dilemmas and the present or future (Yakich 2007). These poems prompt 'revision, realization or new action based on the past' and their present can be marked by 'resignation, epiphany, rejection, lament, self-consciousness [or] foresight' (Yakich 2007: 61). Verb tenses and markers of time indicate these shifts and these sorts of poems have at their heart 'a desire to give order and come to terms with experience and memory' (Yakich 2007: 72). Recognising this can help the poet as she works and reworks the text over time, polishing it as she would with any text that really matters to her.

The poem in this chapter may not yet have found its point of order, and this poet has not come to terms with the experiences within it and the emotions they evoke. A poem can take some time to settle and to provoke. Discussing a data poem that she has revised again and again over eight years, Cahnmann-Taylor acknowledges the time and care such work requires, as the poet who is also a scholar: 'cannot rush the poem or insist the poem be "about" the subject of our research. Poems take on their own natures and timelines and when rushed can result in an aesthetic that feels forced and either over or underwritten' (Cahnmann-Taylor 2009: 27). The poem in this chapter, despite the time frame marked throughout it, was written in several furious bursts with long periods of gestation between them, and multiple revisions. The impasse of the ending in this version came when shocking events exceeded the traces of them in this text and the (im)possibilities of the task the poet had set herself, for the moment, became apparent. Humility and tentativeness are conditions of possibility for the art of poetry, as perhaps they might be more often in any sort of research.

References

Bachelard, G. (1964) *The Poetics of Space*, New York: Beacon Books.

Cahnmann-Taylor, M. (2009) 'The craft, practice and possibility of poetry in educational research', in M. Prendergast, C. Leggo and P. Sameshima (eds) *Poetic Inquiry: Vibrant Voices in the Social Sciences*, Rotterdam: Sense Publishers, 13–29.

Cahnmann-Taylor, M. and Siegesmund, R. (2008) *Arts-Based Research in Education: Foundations for Practice*, London: Routledge.

Camden-Pratt, C. (2008) 'Social ecology and creative pedagogy: Using creative arts and critical thinking in co-creating and sustaining ecological learning webs in university pedagogies', *Transnational Curriculum Inquiry*, 5(1), http://nitinat.library.ubc.ca/ojs/index.php/tci, accessed 31 July 2010.

Cameron, J. (2003) 'Educating for place responsiveness: An Australian perspective on ethical practice', *Ethics, Place and Environment*, 6(2): 99–115.

Foucault, M. (1986) 'Of other spaces', *Diacritics* 16(1): 22–7.

Gannon, S. (2009) 'Writing poetry in/to place', in M. Prendergast, C. Leggo and P. Sameshima (eds) *Poetic Inquiry: Vibrant Voices in the Social Science*, Rotterdam: Sense Publishers, 209–18.

Hill, S. (2003) 'Autonomy, mutualistic relationships, sense of place and conscious caring: A hopeful view of the present and future', in J. Cameron (ed.) *Changing Places: Re-imagining Australia*, Sydney: Longueville, 180–96.

Hill, S, Wilson, S. and Watson, K. (2004) 'Learning ecology. A new approach to learning and ecological consciousness', in E. O'Sullivan and M. Taylor (eds) *Learning Toward an Ecological Consciousness*, New York: Palgrave MacMillan, 47–64.

Massey, D.B. (2005) *For Space*, Thousand Oaks, CA: Sage.

Neilsen, L. (2004) 'Learning to listen: Data as poetry: Poetry as data', *Journal of Critical Inquiry into Curriculum and Instruction*, 5(2): 40–2.

Prendergast, M. (2009) 'Introduction: The phenomena of poetry in research: "Poem is what?" Poetic inquiry in qualitative social science research', in M. Prendergast, C. Leggo and P. Sameshima (eds) *Poetic Inquiry: Vibrant Voices in the Social Sciences*, Rotterdam: Sense Publishers, xix–xliii.

Richardson, L. (1997) *Fields of Play: Constructing an Academic Life*, New Brunswick, NJ: Rutgers University Press

Squire, C., Andrews, M. and Ta.mboukou, M. (2008) 'Introduction: What is narrative research?', in M. Andrews, C. Squire and M. Tamboukou (eds) *Doing Narrative Research*, London: Sage, 1–21.

Wright, D. (2005) 'Embodying, emotioning, expressing learning', *Reflective Practice*, 6(1): 85–93.

Yakich, M. (2007) 'The retrospective-prospective structure', in M. Theune (ed.) *Structure and surprise. Engaging Poetic Turn*, New York: Teachers and Writers Collaborative, 61–82.

INDEX

BOOKS FOR A MORE CREATIVE, PEACEFUL AND SUSTAINABLE WORLD

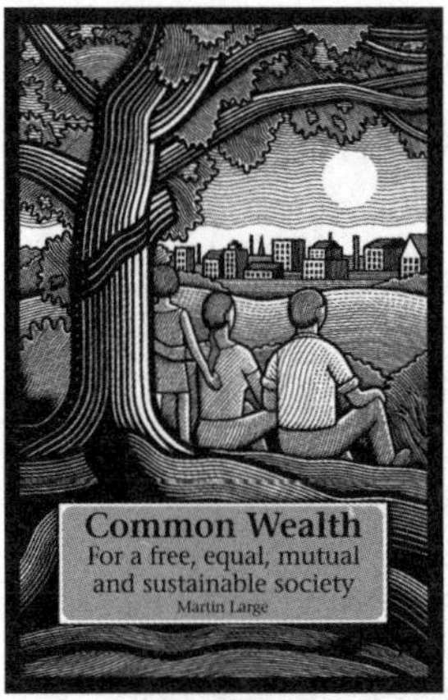

Common Wealth

For a free, equal, mutual and sustainable society

Martin Large

Just when 'the market' nearly took over all areas of life, the credit, climate and democratic crunches came along, challenging us to rebuild a society that works well for all. Martin Largc asks, 'How can we build a more free, equal, mutual and sustainable society?' He argues that the current 'market state', enables a massive wealth transfer to the elite. This is now accelerated by huge handouts to the banks and the privatization of public assets at cut prices. *Common Wealth* shows how the crunches we face can be tackled practically by:

- Re-drawing the boundaries between business, government and civil society
- Pushing back the market from government and public services
- Community land trusteeship
- Mutualising the financial sector and the utilities
- Growing social business by putting capital and business into trusteeship as with the Co-op and John Lewis
- Partnerships between civil society, government and business for sustainable development
- Freeing education from bureaucracy and children from commercialism

'What could be more important to our time than re-establishing the fundamentals for sustainable common wealth? This most timely book, written by one of the leading thinkers in the field, sets out a clear agenda that challenges all of us to act.'
Dr Neil Ravenscroft, Professor of Land Economy, University of Brighton

'In his masterly new book, Martin Large identifies land value taxation and a citizens income as among the measures that will bring the changes about. Please read it if you care about the future of our species.'
James Robertson, co-founder of the New Economics Foundation

'A rest from the insanity of the day – reading *Common Wealth* by Martin Large'
Richard Murphy, Tax Justice Blog, February 15th, 2011

306pp; 234mm × 156mm; ISBN 978-1-903458-98-3; hb

BOOKS FOR A MORE CREATIVE, PEACEFUL AND SUSTAINABLE WORLD

Confronting Conflict

A toolkit for handling conflict

Friedrich Glasl

Conflict costs! When tensions and differences are ignored they grow into conflicts, injuring relationships and organisations. So, how can we confront conflict successfully? Dr Friedrich Glasl has worked with conflict resolution in companies, schools and communities for over thirty years, earning him and his techniques enormous respect. *Confronting Conflict* is authoritative and up to date, containing new examples, exercises, theory and techniques.

You can start by assessing the symptoms and causes of conflict, and ask, 'Am I fanning the flames?' And then consider, 'How can I behave constructively rather than attack or avoid others?' Here, you can:

- Analyse conflict symptoms
- Identify the types, causes of conflict, and if it is hot or cold
- See how personal chemistry, structures or environment influence the conflict
- Acknowledge when you have a conflict, understand conflict escalation, how to lessen conflict through changing behaviour, attitudes and perceptions
- Practice developing considerate confrontation, seizing golden moments, strengthening empathy and much, much more

Confronting Conflict is useful for managers, facilitators, management lecturers and professionals such as teachers and community workers, mediators and workers in dispute resolution.

Dr Friedrich Glasl is an authority on conflict, lecturing in conflict at Salzburg University since 1985.

192pp; 216 × 138mm; ISBN 978-1-869890-71-1; pb

BOOKS FOR A MORE CREATIVE, PEACEFUL AND SUSTAINABLE WORLD

Peace Journalism

Jake Lynch and Annabel McGoldrick

Peace Journalism explains how most coverage of conflict unwittingly fuels further violence, and proposes workable options to give peace a chance. Here are:

Topical case studies including Iraq and 'the war on terrorism' supported by theory, analysis, archive material and photographs to cover:

- A comparison of War Journalism and Peace Journalism
- How the reporting of war, violence and terror can be made more accurate and more useful
- Practical tools and exercises for analysing and reporting the most important war stories of our time

Professional journalists Jake Lynch and Annabel McGoldrick draw on thirty years' experience reporting for the BBC, ITV, Sky News, the London Independent and ABC Australia. They teach Peace Journalism at the University of Sydney. They have led training workshops for editors and reporters in many countries, including the UK, USA, Indonesia, the Philippines, Nepal, the Middle East and the Caucasus.

'Elegantly written, often humorous, always encyclopaedic – the most refreshing and constructive analysis of media practice for years.'

Stuart Rees, Professor Emeritus and Director,
Centre for Peace and Conflict Studies, University of Sydney

'An indispensable training tool for journalists living and working in the midst of violent conflict.'

Carolyn Arguillas, editor, Mindanews, the Philippines

288pp; 246 × 189mm; ISBN 978-1-903458-50-1; pb

Hawthorn Press books are available in Australia from:

Footprint Books Pty Ltd
1/6A Prosperity Parade, Warriewood, NSW 2102
Tel : (+61) 02 9997 3973 Fax : (+61) 02 9997 3185

Study Social Ecology at University of Western Sydney

Studying Social Ecology I experienced an inspirational and personally relevant way of 'doing' and 'being' education that I had not previously experienced in higher education or even secondary education. (Student 2010)

Social Ecology is offered at the University of Western Sydney within the School of Education. Students are able to study a Master of Education (Social Ecology). In 2011–2012 the undergraduate Social Ecology major is under review and listed units are changing. Undergraduate teacher education students can select two or three Social Ecology units which are listed in the compulsory Education Studies Major.

Master of Education (Social Ecology)

At the postgraduate level, our Master of Education (Social Ecology) explores the dynamic interrelationships between the personal, social, environmental and 'spiritual' domains. We acknowledge that everything we do as individuals affects others: our shared communities and environments. We regard ourselves as being influential parts of the systems (local to global) in which we live, and needing to take responsibility for our roles. Key themes include transformative learning and leadership, applied imagination and creativity, environmental education, social action and advocacy, eco-psychology, eco-spirituality, sense of place and the emergent characteristics of complexity, systemics and ecological thinking. Graduates may develop skills in the areas of environmental education, environmental advocacy, ecological systems, environmental futures, critical and transformative learning and leadership and other areas of interest. Students may also choose a variety of other areas of research and independent study under the guidance of experienced academic staff.

For more information see: http://handbook.uws.edu.au/hbook/course.aspx?course=1683.2

Social Ecology at the Undergraduate Level

At the undergraduate level, Social Ecology units explore the rich diversity of relationships between the individual, society and environment. Social Ecology unites theory and practice, social sciences, arts and physical sciences. It provides a transdisciplinary approach to learning and research emphasising reflective thinking, criticality, creativity, aesthetics, participation, communication and ways for enabling transformative change. Students develop skills in ecologically informed analysis, social and environmental action, transformative education and research, with an emphasis on sustainability, social change and leadership.

See: http://handbook.uws.edu.au/hbook/unitset.aspx?unitset=M1015.1 (Major – Social Ecology)
See: http://handbook.uws.edu.au/hbook/unitset.aspx?unitset=M1023.1 (Education Studies Major)

Staff Contact Details

Dr Carol Birrell c.birrell@uws.edu.au

Dr Catherine E. Camden Pratt, Course Adviser Education Studies Major
c.camdenpratt@uws.edu.au http://www.uws.edu.au/education/soe/key_people/academic_staff/doctor_catherine_e_camden_pratt

Dr Brenda Dobia b.dobia@uws.edu.au http://www.uws.edu.au/education/soe/key_people/academic_staff/dr_brenda_dobia

Professor Stuart Hill, Foundation Chair of Social Ecology s.hill@uws.edu.au
www.stuartbhill.com

Dr David Wright, Course Adviser Social Ecology david.wright@uws.edu.au http://www.uws.edu.au/education/soe/key_people/academic_staff/david_wright